Core Books in Advanced Mathematics

Methods of Algebra

J. E. Hebborn
Moderator in Mathematics,
University of London School Examinations Board;
Senior Lecturer in Mathematics,
Royal Holloway and Bedford New College,
University of London.

C. Plumpton
Formerly Moderator in Mathematics, University of
London School Examinations Board; and
formerly Reader in Engineering Mathematics,
Queen Mary College, University of London.

Macmillan Education
London and Basingstoke

© J. E. Hebborn and C. Plumpton 1985

All rights reserved. No reproduction, copy or transmission of this publication may be made without written permission.

No paragraph of this publication may be reproduced, copied or transmitted save with written permission or in accordance with the provisions of the Copyright Act 1956 (as amended).

Any person who does any unauthorised act in relation to this publication may be liable to criminal prosecution and civil claims for damages.

First published 1985

Published by
MACMILLAN EDUCATION LTD
Houndmills, Basingstoke, Hampshire RG21 2XS
and London
Companies and representatives
throughout the world

Printed in Hong Kong

British Library Cataloguing in Publication Data
Hebborn, J. E.
Methods of algebra. —— (Core books in advanced mathematics)
1. Algebra
I. Title II. Plumpton, C. III. Series
512.9   QA152.2
ISBN 0-333-38365-6

Contents

Preface

1 Algebraic functions — 1
Functions, composite functions and inverse functions; Indices, surds and logarithms; The logarithmic and exponential functions; Linear relations

2 Polynomials and rational functions — 17
Polynomials, the remainder theorem and the factor theorem; Rational functions and partial fractions

3 The quadratic function and quadratic equations — 29
Quadratic functions; Quadratic equations

4 Mathematical proof — 42
Some logical concepts; Proof by contradiction (*reductio ad absurdum*); The use of a counter-example; Proof by deduction; Proof by exhaustion; Proof by mathematical induction; Proof of standard results by induction

5 Sequences and series — 54
Sequences; Series; Arithmetic progressions; Arithmetic series; Geometric progressions; Geometric series; Sum of an infinite geometric series; The binomial series; Some other finite series

6 Inequalities — 75
Linear inequalities; Quadratic inequalities; Inequalities involving the modulus sign; More general inequalities in one variable; Inequalities in two variables

Answers — 91
Index — 93

Preface

Advanced level mathematics syllabuses are once again undergoing changes in content and approach following the revolution in the early 1960s which led to the unfortunate dichotomy between 'modern' and 'traditional' mathematics. The current trend in syllabuses for Advanced level mathematics now being developed and published by many GCE Boards is towards an integrated approach, taking the best of the topics and approaches of modern and traditional mathematics, in an attempt to create a realistic examination target through syllabuses which are maximal for examining and minimal for teaching. In addition, resulting from a number of initiatives, core syllabuses are being developed for Advanced level mathematics consisting of techniques of pure mathematics as taught in schools and colleges at this level.

The concept of a core can be used in several ways, one of which is mentioned above, namely the idea of a core syllabus to which options such as theoretical mechanics, further pure mathematics and statistics can be added. The books in this series are core books involving a different use of the core idea. They are books on a range of topics, each of which is central to the study of Advanced level mathematics, which together cover the main areas of any single-subject mathematics syllabus at Advanced level.

Particularly at times when economic conditions make the problems of acquiring comprehensive textbooks giving complete syllabus coverage acute, schools and colleges and individual students can collect as many of the core books as they need to supplement books they already have, so that the most recent syllabuses of, for example, the London, Cambridge, AEB and JMB GCE Boards can be covered at minimum expense. Alternatively, of course, the whole set of core books gives complete syllabus coverage of single-subject Advanced level mathematics syllabuses.

The aim of each book is to develop a major topic of the single-subject syllabuses, giving essential book work, worked examples and numerous exercises arising from the authors' vast experience of examining at this level. Thus, the core books, as well as being suitable for use in either of the above ways, are ideal for supplementing comprehensive textbooks by providing more examples and exercises, so necessary for the preparation and revision for examinations.

The ability to carry out basic algebraic manipulations accurately and quickly is essential for Advanced level mathematics and is the key to success in many other aspects of mathematics. In this particular book we cover the algebraic techniques essential for the core syllabus of pure mathematics now being included by GCE Examining Boards in Advanced level syllabuses. The many worked examples illustrating the various techniques employed form an essential part of this book and are intended to ensure that the conscientious student acquires a mastery of manipulative processes involving functions (algebraic and transcendental), indices, surds, polynomials, quadratic functions and equations, sequences and series, and inequalities. In addition, a chapter on mathematical proof enumerates the different types of mathematical proof expected to be known at this level and contains many algebraic illustrations.

The examples and exercises throughout the book are illustrative of those now being set in single-subject Advanced level mathematics by the GCE examining boards.

John E. Hebborn
Charles Plumpton

1 Algebraic functions

1.1 Functions, composite functions and inverse functions

A *function* is a mapping which associates with each element of one set A a unique element of another set B. The set A is called the *domain* of the function and the set B is called the *codomain* of the function. Not every element of the codomain need have a corresponding element in the domain, but every element of the domain must correspond to some element in the codomain.

The actual elements of the set B which are images of the elements of the domain are called the *range* (or range set) of the function.

We will only be concerned with functions which map a real variable x to a real variable y:

$$f: x \mapsto y = f(x).$$

We usually refer to x as the *independent variable* and to y as the *dependent variable*.

There are two important ways of representing a function f from $\mathbb{R}$ to $\mathbb{R}$ diagrammatically.

(i) The domain and codomain are represented by two parallel number lines with an arrow from x to its image $y = f(x)$ (see Fig. 1.1).

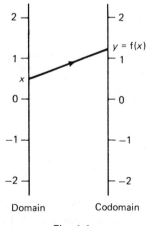
Fig. 1.1

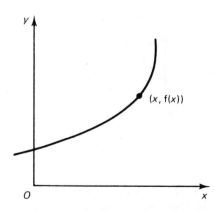
Fig. 1.2

(ii) Each element x of the domain and its image f(x) form an ordered pair [x, f(x)]. This ordered pair can be represented by the coordinates (x, y) of a point in a cartesian plane. The set of all such points is called the *graph* of the function, and the relation $y = $ f(x) is called the equation of the graph or curve (see Fig. 1.2).

A function f is said to be an *even* function if

$$f(x) = f(-x) \text{ for all values of } x.$$

A function f is said to be an *odd* function if

$$f(-x) = -f(x) \text{ for all values of } x.$$

Example 1
(a) The mapping $x \to x^{1/2}$ does not define a function, since, for a given real number, there is not a unique element of the image set corresponding to it. (For each real value of x there are two elements of the image set corresponding to it, $+\sqrt{x}$ and $-\sqrt{x}$.)
(b) The mapping $x \to x + 2$ is a function. The domain is the set $\mathbb{R}$ of real numbers, or some subset of $\mathbb{R}$. The codomain is also the set of real numbers. It may be represented as in Fig. 1.3(a) or Fig. 1.3(b).

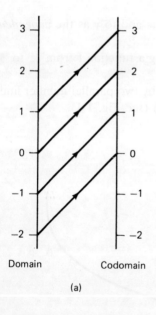
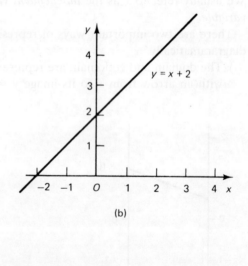

Fig. 1.3

(c) The mapping $x \to x^2$ is a function. The domain is $\mathbb{R}$ and the codomain is $\mathbb{R}$. The range is the set $\{x: x \in \mathbb{R}, x \geq 0\}$. It may be represented as in Fig. 1.4(a) or Fig. 1.4(b).

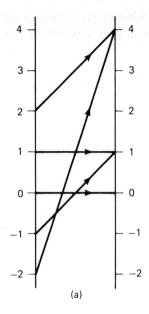
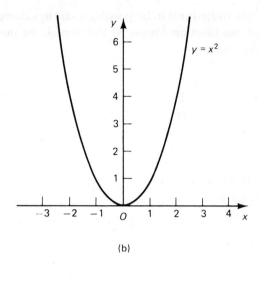

Fig. 1.4

Example 2 Determine whether the functions f, g, h, where (a) $f(x) = 3x^2$, (b) $g(x) = x - 2x^3$, (c) $h(x) = (x - 2)/(x + 2)$, are odd, even or neither.

(a) $f(-x) = 3(-x)^2 = 3x^2 = f(x) \Rightarrow$ f is even.
(b) $g(-x) = (-x) - 2(-x)^3 = -x + 2x^3 = -[x - 2x^3] = -g(x) \Rightarrow$ g is odd.
(c) $h(-x) = (-x - 2)/(-x + 2)$ and this is neither $h(x)$ nor $-h(x)$. Hence, $h(x)$ is neither odd nor even.

Composition of functions

It is possible, under certain circumstances, to combine functions. This is best illustrated by a simple example. Suppose

$$f: x \mapsto x^2$$

and

$$g: x \mapsto x + 1.$$

Then the composite function, written gf (note the order), means 'square x and then add 1' — i.e. f first, followed by g:

$$gf: x \mapsto x^2 + 1.$$

This is not the same as fg, which means 'add 1 to x and then square':

$$fg: x \mapsto (x + 1)^2.$$

Algebraic functions 3

In order for it to be possible to form a composite function gf, the image set of the function f must be the domain of the function g or a subset of this domain.

Example 3 If f: $x \mapsto x^2 + 5$, g: $x \mapsto x + 2$, then
(a) gf: $x \mapsto (x^2 + 5) + 2 \equiv x^2 + 7$,
(b) fg: $x \mapsto (x + 2)^2 + 5 \equiv x^2 + 4x + 4 + 5 = x^2 + 4x + 9$,
(c) ff: $x \mapsto (x^2 + 5)^2 + 5 \equiv x^4 + 10x^2 + 25 + 5$
$\qquad\qquad\qquad\quad = x^4 + 10x^2 + 30$,
(d) gg: $x \mapsto (x + 2) + 2 \equiv x + 4$.
Note that
$$[f(x)]^2 = (x^2 + 5)^2 \equiv x^4 + 10x^2 + 25$$
and this is not the result of applying the mapping ff to x.

Identities and equations

The reader will note that in the above example we have used the symbol $\equiv$, which is to be read as 'is identical with'. We use this symbol when we have an algebraic relation which is true for all values of the variable x. For example,
$$(x + 2)^2 \equiv x^2 + 4x + 4.$$
Such an algebraic relation is called an *identity*.
On the other hand, the expression
$$(x + 2)^2 = 9x - 2$$
is only true when $x = 2$ or $x = 3$. An algebraic relation which is only true for a particular set of values of x is called an *equation*.

Inverse functions

If for a given function f a function g can be found such that
$$\text{gf: } x \mapsto x$$
and also
$$\text{fg: } x \mapsto x,$$
then g is denoted by f^{-1} and is said to be the *inverse* of f.
In general, for a function f to have an inverse f^{-1} the following conditions must be satisfied:
(i) the range of f = the codomain of f,
(ii) $x_1 = x_2 \Rightarrow f(x_1) = f(x_2)$,
(iii) $f(x_1) = f(x_2) \Rightarrow x_1 = x_2$,
 $\}$ for all x_1 and x_2 in the domain of f.

Example 4
(a) f: $x \mapsto 3x + 1$.
Let $y = 3x + 1$, so that $x = (y - 1)/3$. Then f: $(y - 1)/3 \mapsto y$
$$\Rightarrow f^{-1}: y \mapsto (y - 1)/3 \quad \text{or} \quad f^{-1}: x \mapsto (x - 1)/3,$$
with domain and range $\mathbb{R}$.

(b) f: $x \mapsto \dfrac{1}{x}$.

Let $y = \dfrac{1}{x} \Rightarrow x = \dfrac{1}{y} \Rightarrow$ f: $\dfrac{1}{y} \mapsto y$.

Therefore, $f^{-1}: y \mapsto \dfrac{1}{y}$ or $f^{-1}: x \mapsto \dfrac{1}{x}$.

Hence, f is its own inverse.

The above examples suggest the following rule for obtaining the inverse.

If $y = f(x)$ is the equation of the graph for any function f which has an inverse, then (i) interchange x and y so that $x = f(y)$, (ii) solve, if possible, for y in terms of x.

The result gives the inverse function f^{-1} if it exists.

(c) Consider the function f: $x \mapsto x^2$ with the domain $\mathbb{R}$. Clearly, both 2 and -2 map to 4. As it is not possible to say uniquely which point maps to 4, this function does not have an inverse.

However, if we restrict the domain to $\mathbb{R}^+$, the set of positive real numbers, then f maps onto the codomain $\mathbb{R}^+$ and f does have an inverse:
$$f^{-1}: x \mapsto \sqrt{x}.$$

[Note we use $\sqrt{x}$ to denote the positive square root of x.]

(d) If f: $x \mapsto x + 1$ and
g: $x \mapsto x^2$,
then
fg: $x \mapsto x^2 + 1$.

The inverse is obtained by solving $y^2 + 1 = x$, which gives $y = \pm\sqrt{(x - 1)}$. For $(fg)^{-1}$ to be a function we must restrict the new domain to $x \geq 1$ and the range to, say, the positive square root of $(x - 1)$. Then
$$(fg)^{-1}: x \mapsto +\sqrt{(x - 1)} \quad \text{for } x \geq 1.$$

Notice that
$$f^{-1}: x \mapsto x - 1,$$
$$g^{-1}: x \mapsto +\sqrt{x}$$
and
$$g^{-1}f^{-1}: x \mapsto +\sqrt{(x - 1)} \quad \text{for } x \geq 1.$$

Hence, with above restrictions on domain and range
$$(fg)^{-1} = g^{-1}f^{-1}.$$

Note the order, which may at first seem strange. It does, however, agree with practical experience — 'drive the car into the garage and then close the door' but the inverse is *first* open the door and *then* drive the car out'.

The general result is: for all functions f, g, ..., p, q such that the inverse function $(fg \ldots pq)^{-1}$ exists

$$(fg \ldots pq)^{-1} = q^{-1}p^{-1} \ldots g^{-1}f^{-1}.$$

This may be proved by induction (see page 46).

1.2 Indices, surds and logarithms

Indices
The three basic rules of indices are:
  (i) To *multiply* powers of the same base *add indices*:
$$a^m \times a^n \equiv a^{m+n}.$$
  (ii) To *divide* powers of the same base *subtract indices*:
$$a^m \div a^n \equiv a^{m-n}.$$
  (iii) To raise a power of a base to a second index *multiply the indices*:
$$(a^m)^n \equiv a^{mn}.$$

These rules apply for $m, n \in \overline{\mathbb{Z}^+}$. If we apply them to $m, n \in \mathbb{Q}$, we require the following interpretations:

Negative index, $a^{-m} \equiv \dfrac{1}{a^m}$ for $m > 0$.

Zero index, $a^0 \equiv 1$.
Simple fractional index, $a^{1/m} \equiv \sqrt[m]{a}$ for $m > 0$.
Rational indices, $a^{n/m} \equiv (\sqrt[m]{a})^n$ for $m > 0$.

Example 5
(a) $\left(\dfrac{25}{36}\right)^{3/2} = \left[\sqrt{\left(\dfrac{25}{36}\right)}\right]^3 = \left(\dfrac{5}{6}\right)^3 = \dfrac{125}{216}.$

(b) $(27)^{2/3} = (\sqrt[3]{27})^2 = 3^2 = 9.$

(c) $(16)^{3/4} \times (8)^{-1/3} = \dfrac{(\sqrt[4]{16})^3}{\sqrt[3]{8}} = \dfrac{2^3}{2} = 4.$

(d) $4^{-1/2} + \left(\dfrac{1}{81}\right)^{1/4} - (64)^{-1/3} = \dfrac{1}{4^{1/2}} + \dfrac{1}{(81)^{1/4}} - \dfrac{1}{(64)^{1/3}}$

$= \dfrac{1}{2} + \dfrac{1}{3} - \dfrac{1}{4} = \dfrac{6 + 4 - 3}{12} = \dfrac{7}{12}.$

Example 6
(a) $\dfrac{y^{1/6} y^{-1/3}}{y^{1/4}} = y^{1/6 - 1/3 - 1/4} = y^{(2-4-3)/12} = y^{-5/12}.$

(b) $\dfrac{x(x+1)^{1/2} - (x+1)^{-1/2}}{x^3} = \dfrac{x(x+1) - 1}{x^3(x+1)^{1/2}} = \dfrac{x^2 + x - 1}{x^3(x+1)^{1/2}}.$

(c) $\dfrac{\sqrt{x}\sqrt{x^3}}{x^{-4}} = \sqrt{x^4}.x^4 = x^2.x^4 = x^6.$

Surds

Certain numbers of the set of numbers defined by $\sqrt{x}$, for $x \in \mathbb{Z}^+$, have exact numerical values, e.g. $\sqrt{1}, \sqrt{4}, \sqrt{9}, \sqrt{16}, \sqrt{25}, \ldots$. Other numbers of the set, e.g. $\sqrt{2}, \sqrt{3}, \sqrt{5}$, cannot be written as numerically exact quantities, and it is frequently more convenient to leave them in their basic $\sqrt{x}$ form. As such, these numbers are called *surds*, and expressions involving them are called *surd expressions*. It is desirable to write such expressions in their simplest form. The examples below indicate how one should proceed. Two basic rules, derived from the rules of indices, are employed:

$$(\sqrt{x}) \times (\sqrt{y}) = \sqrt{(xy)} \quad [\text{from } x^{1/2}y^{1/2} = (xy)^{1/2}],$$

and

$$\dfrac{\sqrt{x}}{\sqrt{y}} = \sqrt{\left(\dfrac{x}{y}\right)} \quad \left[\text{from } \dfrac{x^{1/2}}{y^{1/2}} = \left(\dfrac{x}{y}\right)^{1/2}\right].$$

Example 7 Reduce the following surd expressions to their simplest form:
(a) $\tfrac{1}{3}\sqrt{18}$, (b) $2\sqrt{2}(\sqrt{32} + \sqrt{2})$, (c) $(2\sqrt{5} + 1)(3\sqrt{5} + 2)$,
(d) $\sqrt{20} + \sqrt{45} - \sqrt{80} + \sqrt{5}$.

(a) $\tfrac{1}{3}\sqrt{18} = \tfrac{1}{3}\sqrt{(9 \times 2)} = \tfrac{1}{3}3\sqrt{2} = \sqrt{2}.$
(b) $2\sqrt{2}(\sqrt{32} + \sqrt{2}) = 2\sqrt{2}(4\sqrt{2} + \sqrt{2}) = 2\sqrt{2}(5\sqrt{2}) = 20.$
(c) $(2\sqrt{5} + 1)(3\sqrt{5} + 2) = 6 \times 5 + 7\sqrt{5} + 2 = 32 + 7\sqrt{5}.$
(d) $\sqrt{20} + \sqrt{45} - \sqrt{80} + \sqrt{5} = 2\sqrt{5} + 3\sqrt{5} - 4\sqrt{5} + \sqrt{5} = 2\sqrt{5}.$

Example 8 We remove the surds from the denominator in the following surd expressions (this process is called rationalising the denominator):

(a) $\dfrac{2}{\sqrt{18}} = \dfrac{2}{3\sqrt{2}} = \dfrac{2\sqrt{2}}{3.2} = \dfrac{\sqrt{2}}{3}.$

(b) $\dfrac{1}{\sqrt{2} - 1}.$

We make use here of the identity

$$(x - y)(x + y) \equiv x^2 - y^2.$$

Multiplying numerator and denominator by $\sqrt{2} + 1$, we obtain

$$\dfrac{1}{\sqrt{2} - 1} \cdot \dfrac{\sqrt{2} + 1}{\sqrt{2} + 1} = \dfrac{\sqrt{2} + 1}{2 - 1} = \sqrt{2} + 1.$$

Algebraic functions 7

(c) $\dfrac{2\sqrt{3}}{2\sqrt{5} - \sqrt{3}} = \dfrac{2\sqrt{3}}{2\sqrt{5} - \sqrt{3}} \cdot \dfrac{(2\sqrt{5} + \sqrt{3})}{(2\sqrt{5} + \sqrt{3})} = \dfrac{4\sqrt{15} + 2.3}{20 - 3}$
$= \dfrac{4\sqrt{15} + 6}{17}.$

(d) $\dfrac{\sqrt{3}}{\sqrt{3} - 1} + \dfrac{\sqrt{3}}{\sqrt{3} + 1}.$

Rationalising each term, we obtain

$\dfrac{\sqrt{3}(\sqrt{3} + 1)}{(\sqrt{3} - 1)(\sqrt{3} + 1)} + \dfrac{\sqrt{3}(\sqrt{3} - 1)}{(\sqrt{3} + 1)(\sqrt{3} - 1)} = \dfrac{3 + \sqrt{3}}{2} + \dfrac{3 - \sqrt{3}}{2} = 3.$

Logarithms

The term 'logarithm' is an alternative word for an index or power of a given positive number base. For example, since $2^3 = 8$, we define the index 3 to be the logarithm of 8 to the base 2 and write

$$3 = \log_2 8.$$

Further, using the rule of negative indices, $(\tfrac{1}{3})^{-2} = 9$ and we may write

$$\log_{1/3} 9 = -2.$$

The base of a logarithm may be any positive number. The tables of common logarithms, which are sometimes used for calculations, have base 10. It is usual, when using common logarithms, to omit the base 10, and write log or lg. In general,

$$a^x = y \Leftrightarrow x = \log_a y.$$

Example 9 Express in logarithmic form (a) $5^2 = 25$, (b) $2^5 = 32$, (c) $6^0 = 1$, (d) $(\tfrac{1}{5})^{-3} = 125$, (e) $10^{-2} = 0 \cdot 01$.

(a) $5^2 = 25 \Rightarrow 2 = \log_5 25.$
(b) $2^5 = 32 \Rightarrow 5 = \log_2 32.$
(c) $6^0 = 1 \Rightarrow 0 = \log_6 1.$
(d) $(\tfrac{1}{5})^{-3} = 125 \Rightarrow -3 = \log_{1/5} 125.$
(e) $10^{-2} = 0 \cdot 01 \Rightarrow -2 = \log_{10} 0 \cdot 01.$

Example 10 Evaluate (a) $\log_4 64$, (b) $\log_{10} 0 \cdot 001$, (c) $\log_{1/2} 4$.

(a) Let $x = \log_4 64 \Rightarrow 4^x = 64 = 4^3.$
Comparing indices gives $x = 3.$
(b) Let $y = \log_{10} 0 \cdot 001 \Rightarrow 10^y = 0 \cdot 001 = 10^{-3}.$
Therefore,
$$y = -3.$$

(c) Let $z = \log_{1/2} 4 \Rightarrow (\frac{1}{2})^z = 4 = 2^2 = (\frac{1}{2})^{-2}$.
Therefore,
$$z = -2.$$

Example 11 Express in index form (a) $\log_5 125 = 3$, (b) $\log_{10} 100 = 2$, (c) $\log_{36} 6 = \frac{1}{2}$, (d) $\log_a 1 = 0$, (e) $\log_x y = z$.

(a) $\log_5 125 = 3 \Rightarrow 5^3 = 125$.
(b) $\log_{10} 100 = 2 \Rightarrow 10^2 = 100$.
(c) $\log_{36} 6 = \frac{1}{2} \Rightarrow 36^{1/2} = 6$.
(d) $\log_a 1 = 0 \Rightarrow a^0 = 1$.
(e) $\log_x y = z \Rightarrow x^z = y$.

Rules of logarithms

(1) Addition of logarithms
If $\log_a x = m$ and $\log_a y = n$, then $x = a^m$, $y = a^n$
and
$$xy = a^m \times a^n = a^{m+n}$$
$$\Rightarrow \log_a(xy) = m + n.$$
Therefore,
$$\log_a(xy) \equiv \log_a x + \log_a y.$$

(2) Subtraction of logarithms
Similarly,
$$\frac{x}{y} = \frac{a^m}{a^n} = a^{m-n}.$$
Hence,
$$\log_a\left(\frac{x}{y}\right) \equiv \log_a x - \log_a y.$$

(3) Logarithm of powers of numbers
$$x^p \equiv (a^m)^p = a^{mp}.$$
Hence,
$$\log_a x^p = mp = p \log_a x.$$
Therefore,
$$\log_a x^p = p \log_a x.$$

(4) Change of base of a logarithm

$$\log_a x \equiv m \Rightarrow a^m = x.$$
$$\log_b x \equiv \log_b(a^m) = m \log_b a$$
$$\Rightarrow \log_b x \equiv (\log_a x) \times (\log_b a),$$

i.e. to change the base of a logarithm of x from a to b we multiply by $\log_b a$.
If in this result we replace x by a, we get

$$\log_b a \equiv (\log_a a) \times (\log_b a).$$

Dividing by $\log_b a$, given that this is non-zero,

$$\Rightarrow 1 = \log_a a,$$

and, if $x = b$,

$$\log_b b \equiv 1 \equiv (\log_a b) \times (\log_b a).$$

Example 12 Simplify
(a) $4 \log_a 2 + 2 \log_a 3$,
(b) $\log_{10}\left(\dfrac{15}{64}\right) - 2 \log_{10}\left(\dfrac{5}{3}\right) + \log_{10}\left(\dfrac{16}{9}\right)$,
(c) $3 \log_a x + 2 \log_a y - \log_a z$.

(a) $4 \log_a 2 + 2 \log_a 3 = \log_a 2^4 + \log_a 3^2$ by rule (3),
    $= \log_a 16 + \log_a 9 = \log_a 144$ by rule (1).
(b) $\log_{10}\left(\dfrac{15}{64}\right) - 2 \log_{10}\left(\dfrac{5}{3}\right) + \log_{10}\left(\dfrac{16}{9}\right)$
    $= \log_{10}\left(\dfrac{15}{64}\right) - \log_{10}\left(\dfrac{5^2}{3^2}\right) + \log_{10}\left(\dfrac{16}{9}\right)$ by rule (3),
    $= \log_{10}\left(\dfrac{15}{64} \times \dfrac{9}{25} \times \dfrac{16}{9}\right)$ by rules (1) and (2),
    $= \log_{10}\left(\dfrac{3}{20}\right).$
(c) $3 \log_a x + 2 \log_a y - \log_a z$
    $= \log_a x^3 + \log_a y^2 - \log_a z$ by rule (3),
    $= \log_a\left(\dfrac{x^3 y^3}{z}\right)$ by rules (1) and (2).

Example 13 Using only the rules of logarithms, find the value of $\log_3 125 \times \log_5 9$.

$\log_3 125 \times \log_5 9 = \log_3 5^3 \times \log_5 3^2$
$= 3 \log_3 5 \times 2 \log_5 3 = 6(\log_3 5)(\log_5 3)$
$= 6 \log_5 5 = 6.$

1.3 The logarithmic and exponential functions

In calculus logarithms to a particular base are used. The logarithms are called *natural logarithms* and the base is denoted by e. The number $e \approx 2.7182818$ and is defined as that base for which the function $\log_e x$ has unit gradient at $x = 1$. The standard notation is

$$\log_e x = \ln x.$$

The function $\ln x$ has domain $\mathbb{R}^+$ and is called the *logarithmic function*. The range of $\ln x$ is $\mathbb{R}$. Its inverse function is known as the *exponential function*, $\exp x$, which is denoted by e^x. The domain of the exponential function is $\mathbb{R}$ and its range is $\mathbb{R}^+$. To summarise,

$$y = \ln x \Leftrightarrow x = e^y.$$

The graphs of $y = e^x$ and $y = \ln x$ are given in Fig. 1.5. Note that, since e^x and $\ln x$ are inverse functions, their graphs are mirror images of each other in the line $y = x$ (shown dotted).

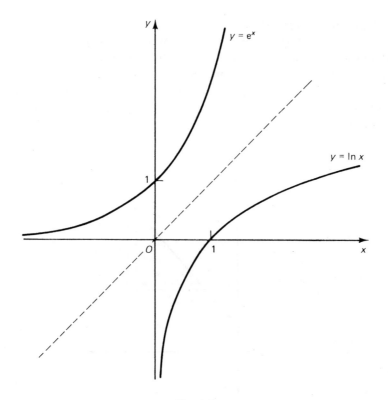

Fig. 1.5

Example 14 Simplify

$$\frac{e^{2x} - \ln e}{e^x + e^{\ln 1}}.$$

We notice that $\ln e = 1$, and, since $\ln 1 = 0$, we have $e^{\ln 1} = e^0 = 1$. Therefore,

$$\frac{e^{2x} - \ln e}{e^x + e^{\ln 1}} \equiv \frac{e^{2x} - 1}{e^x + 1}.$$

Since

$$e^{2x} \equiv (e^x)^2$$

we may write

$$e^{2x} - 1 \equiv (e^x + 1)(e^x - 1)$$

and so

$$\frac{e^{2x} - \ln e}{e^x + e^{\ln 1}} \equiv \frac{(e^x + 1)(e^x - 1)}{e^x + 1} \equiv e^x - 1.$$

1.4 Linear relations

An equation of the form $px + qy + r = 0$, where p and q are non-zero constants, in which x and y occur but y^2, x^2, yx and other products of x and y do not occur, is called a linear equation. The variables x and y are said to satisfy a *linear relation*. We say that a linear relationship exists between x and y.

It is usual to divide this relation by q and rearrange the equation in the standard form

$$y = mx + c.$$

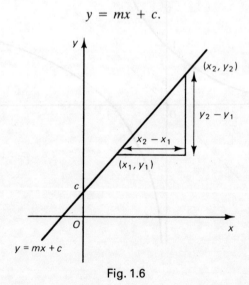

Fig. 1.6

When the equation is written in this form, the constants m and c have a simple interpretation which is easily obtained by drawing the graph of y against x.

When $x = 0$, $y = c \Rightarrow c$ is the intercept on the y-axis. If (x_1, y_1) and (x_2, y_2) are any two points on the line, then

$$\left.\begin{array}{l} y_1 = mx_1 + c \\ y_2 = mx_2 + c \end{array}\right\} \Rightarrow y_2 - y_1 = m(x_2 - x_1) \Rightarrow m = \frac{y_2 - y_1}{x_2 - x_1}.$$

Hence, m is the gradient of the line. Care should be taken with the sign of m. A negative m indicates that the line makes an obtuse angle with Ox. (See Example 15 below.)

Example 15 Find the gradient and the intercepts on the x- and y-axes of the line $2x + 4y = 6$.

Written in standard form, the line becomes $y = -\frac{1}{2}x + \frac{3}{2}$. From the above $m = -\frac{1}{2}$ and $c = \frac{3}{2}$. Hence, the gradient is $-\frac{1}{2}$ and the intercept on the y-axis is $\frac{3}{2}$. The graph is shown in Fig. 1.7.

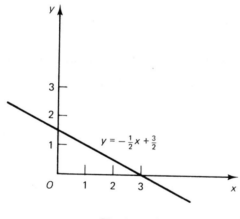

Fig. 1.7

When $y = 0$ we have $-\frac{1}{2}x + \frac{3}{2} = 0 \Rightarrow x = 3$. Hence, the intercept on the x-axis is 3.

Reduction to linear form

Above we have seen that the essential constants in a linear form are easily found graphically. If it is known that two physical variables x and y are related in such a way, then the precise nature of the relationship may be found by:
(i) obtaining some pairs of values (x, y),
(ii) drawing the graph determined by them,
(iii) reading off the gradient and the y intercept.

However, many of the relationships between pairs of physical variables are not linear. We now consider how these may be reduced to a linear form by making suitable changes of variable. The following example illustrates the technique.

Example 16 Show, by suitable change of variables, that the following may be converted to linear relationships: (a) $y = kx^n$, (b) $y = ab^x$, (c) $ax + by = xy$, (d) $y = ax^3 + bx$.

(a) If we take the logarithm of both sides of the equation, we have

$$\log y = \log kx^n = \log k + \log x^n = \log k + n \log x.$$

Defining $Y = \log y$ and $X = \log x$, we obtain

$$Y = \log k + nX,$$

which is a linear relationship between X and Y.

(b) Again taking logarithms, we obtain

$$\log y = \log a + x \log b.$$

If $Y = \log y$ we obtain $Y = \log a + x \log b$, again a linear relationship, this time between Y and x.

(c) Given that $ax + by = xy$, we have, on dividing by xy,

$$\frac{a}{y} + \frac{b}{x} = 1.$$

Setting $Y = \dfrac{1}{y}$ and $X = \dfrac{1}{x}$, we have $aY + bX = 1$, a linear relationship as required.

(d) On dividing $y = ax^3 + bx$ by x, we obtain $\dfrac{y}{x} = ax^2 + b$.

If $Y = \dfrac{y}{x}$ and $X = x^2$, we have $Y = aX + b$, a linear relationship.

Exercise 1

1. State the most extensive possible domain and the corresponding range for each of the following relations to be a function:
    (a) f: $x \mapsto 1 - 2x$, (b) g: $x \mapsto \dfrac{1}{1 + x^2}$, (c) h: $x \mapsto 5x^4 + 2$.
    Give sketches, such as those in Example 1, for each example.
2. Which of the functions in Question 1 are odd, even or neither?
3. If f: $x \mapsto 2x - 1$ and g: $x \mapsto x^3$, express in the form $x \mapsto \ldots$ the functions gf, fg, ff and gg.
4. Find inverses of the following functions, stating a suitable domain for each:
    (a) f: $x \mapsto 5 - 2x$, (b) g: $x \mapsto \dfrac{4}{x - 1}$, (c) h: $x \mapsto (2x + 1)^2$.

5 Suggest suitable domains and codomains for the following relations to define functions with inverse functions:
   (a) $r_1: x \mapsto 16x^2$, (b) $r_2: x \mapsto \dfrac{1}{2+x}$, (c) $r_3: x \mapsto x^2 - 4$.
6 The domain of the function f is the set
$$D = \{x: x \in \mathbb{R}, -2 < x < 3\}.$$
The function f: $D \mapsto$ is defined by
$$f(x) = \begin{cases} 3x - 2 & \text{for } -2 < x \leq 1, \\ x^2 & \text{for } 1 < x \leq 2, \\ 8 - 2x & \text{for } 2 < x < 3. \end{cases}$$
Find the range of this function and sketch a graph of this function. Explain why there is no inverse function to f. Suggest an interval such that f, restricted to this interval, will have an inverse function. Give an expression for the inverse function in this case.

7 Evaluate, without the use of tables or a calculator,
   (a) $9^{-3/2}$, (b) $\left(\dfrac{125}{8}\right)^{2/3}$, (c) $\dfrac{25^{1/2} \times 16^{-1/4}}{36^{3/2}}$, (d) $\left(\dfrac{1}{3}\right)^{-2} \times 4^0$,
   (e) $9^{1/2} + \left(\dfrac{1}{125}\right)^{1/3} - (36)^{-1/2}$.

8 Reduce each of the following expressions to its simplest form:
   (a) $\dfrac{\sqrt{x}\sqrt{x^5}}{x^{-2}}$,
   (b) $\dfrac{y^{1/3} y^{-1/2}}{y^{1/6}}$,
   (c) $(x^5)^2 \div (x^3)^3$,
   (d) $6x^{-3/2} \div 3x^{-1/2}$,
   (e) $\dfrac{(x+1)^{-1/3} - 2(x+1)^{2/3}}{(x+1)^{1/3}}$,
   (f) $x^{-1} + 2x^{-3} - 3x^{-2}$.

9 Reduce the following surd expressions to their simplest form:
   (a) $\sqrt{90}$, (b) $\tfrac{1}{2}\sqrt{48}$, (c) $\sqrt{18} \times \sqrt{50}$, (d) $(2\sqrt{3} - 1)(\sqrt{3} + 2)$,
   (e) $\sqrt{12} + \sqrt{27} - \sqrt{75}$.

10 Express the following surd expressions in a form in which the denominators are rational numbers:
   (a) $\dfrac{3}{\sqrt{27}}$, (b) $\dfrac{\sqrt{3}}{3\sqrt{2} - 2\sqrt{3}}$, (c) $\dfrac{1}{\sqrt{2}-1} - \dfrac{1}{\sqrt{2}+1}$.

11 Express in logarithmic form
   (a) $3^2 = 9$, (b) $8^0 = 1$, (c) $(\tfrac{1}{3})^{-4} = 81$, (d) $a^b = 2$.

12 Evaluate, without use of tables or a calculator,
   (a) $\log_{1/2} 8$, (b) $\log_{10} 0 \cdot 1$, (c) $\log_{16} 64$.

13 Express in index form
   (a) $\log_{10} 10 = 1$, (b) $\log_3 81 = 4$, (c) $\log_{27} 3 = \tfrac{1}{3}$.

14 Simplify
   (a) $\tfrac{1}{2} \log_a 9 + 3 \log_a 2 - 2 \log_a 4$,
   (b) $3 \log_{10} 2 - \tfrac{1}{4} \log_{10} 16 + 2 \log_{10} 5$.

15 If $\log_{10} 2 = a$, show that $\log_8 5 = \left(\dfrac{1-a}{3a}\right)$.

16 Simplify

$$\dfrac{e^{2x} - \ln e}{[e^{(x+1)/2}]^2 + e}.$$

17 For each of the following find a change of variables which will produce a linear relation:
(a) $v = a e^{nu}$, (b) $s = ut + \tfrac{1}{2}ft^2$, (c) $x^k y = a$.

2 Polynomials and rational functions

2.1 Polynomials, the remainder theorem and the factor theorem

An expression of the form

$$a_n x^n + a_{n-1} x^{n-1} + \cdots + a_1 x + a_0 \quad (2.1)$$

where $a_n, a_{n-1}, \ldots, a_1, a_0$ are constants and n is a positive integer, is called a *polynomial*. The highest power of x occurring in the expression defines the *degree* or *order* of the polynomial and the a_is are called the *coefficients*. If $a_n \neq 0$ in Equation (2.1), the polynomial is of degree n and $a_i (i = 1, 2, \ldots, n)$ is the coefficient of x^i. The term not involving x, namely a_0, is called the constant term. It is usual to write the polynomial in a systematic way either in descending powers of x as in Equation (2.1) or in ascending powers of x, when Equation (2.1) becomes

$$a_0 + a_1 x + \cdots + a_{n-1} x^{n-1} + a_n x^n. \quad (2.2)$$

Polynomials of degree 2, 3 and 4 are called *quadratics*, *cubics* and *quartics*, respectively.

Manipulating polynomials

Addition and subtraction
To add or subtract polynomials, we collect together terms of the same degree and combine these, using the distributive law.

Example 1 Given that $P(x) \equiv 2x^3 + 3x^2 + 2$ and $Q(x) \equiv 4x^4 + 3x^3 + 5x + 1$, find $P(x) + Q(x)$ and $P(x) - Q(x)$.

$$\begin{aligned}
P(x) + Q(x) &\equiv (2x^3 + 3x^2 + 2) + (4x^4 + 3x^3 + 5x + 1) \\
&\equiv 4x^4 + (2x^3 + 3x^3) + 3x^2 + 5x + (2 + 1) \\
&\equiv 4x^4 + 5x^3 + 3x^2 + 5x + 3.
\end{aligned}$$

$$\begin{aligned}
P(x) - Q(x) &\equiv (2x^3 + 3x^2 + 2) - (4x^4 + 3x^3 + 5x + 1) \\
&\equiv -4x^4 + (2x^3 - 3x^3) + 3x^2 - 5x + (2 - 1) \\
&\equiv -4x^4 - x^3 + 3x^2 - 5x + 1.
\end{aligned}$$

Multiplication

Also, we now require the distributive property of multiplication over addition. This is illustrated in Examples 2 and 3.

Example 2 Work out

$$x^2(x^2 + x + 1) + x(x^2 + 1)$$

as a polynomial in descending powers of x.

$$x^2(x^2 + x + 1) + x(x^2 + 1)$$
$$\equiv (x^4 + x^3 + x^2) + (x^3 + x)$$
$$\equiv x^4 + (x^3 + x^3) + x^2 + x \equiv x^4 + 2x^3 + x^2 + x.$$

Example 3 Multiply $(x^2 - 2x + 1)$ by $(x^2 + x - 2)$.

$$(x^2 - 2x + 1)(x^2 + x - 2)$$
$$\equiv x^2(x^2 + x - 2) - 2x(x^2 + x - 2) + 1(x^2 + x - 2)$$
$$\equiv (x^4 + x^3 - 2x^2) + (-2x^3 - 2x^2 + 4x) + (x^2 + x - 2)$$
$$\equiv x^4 + (x^3 - 2x^3) + (-2x^2 - 2x^2 + x^2) + (4x + x) - 2$$
$$\equiv x^4 - x^3 - 3x^2 + 5x - 2.$$

The similarity to ordinary long multiplication is easily seen when the working for Example 3 is set out in the following way:

$$
\begin{array}{r}
x^2 - 2x + 1 \\
x^2 + x - 2 \\
\hline
-2x^2 + 4x - 2 \\
x^3 - 2x^2 + x \\
x^4 - 2x^3 + x^2 \\
\hline
x^4 - x^3 - 3x^2 + 5x - 2
\end{array}
$$

i.e. $-2(x^2 - 2x + 1)$
i.e. $x(x^2 - 2x + 1)$
i.e. $x^2(x^2 - 2x + 1)$

It is important to *order* the polynomials before carrying out this process. Notice that in algebra there are minus signs and there is no 'carrying'.

An alternative format, which is useful when programming a computer to handle polynomials, is called the method of detached coefficients. In this format, Example 3 becomes

x^4	x^3	x^2	x	1
		1	−2	1
		1	1	−2
		−2	4	−2
	1	−2	1	
1	−2	1		
1	−1	−3	5	−2

Care must be taken to leave space for missing terms so that like terms may be kept in vertical columns.

Example 4 Multiply $(x^3 + x - 1)(2x - 3)$.

$$
\begin{array}{r}
x^3 + 0x^2 + x - 1 \\
2x - 3 \\
\hline
-3x^3 + 0x^2 - 3x + 3 \\
2x^4 + 0x^3 + 2x^2 - 2x \\
\hline
2x^4 - 3x^3 + 2x^2 - 5x + 3
\end{array}
$$

With some practice it is possible to multiply polynomials mentally. For example, in Example 4 there is only one way that the coefficients of x^4, x^3, x^2 and x^0 can be obtained. The coefficient of x arises from $x(-3)$ and $(-1)2x$.

Division

Since division is the reverse process of multiplication, it is possible to relate the process of division of one polynomial by another to the well-known long division process in arithmetic. Before attempting this, it is important to (i) *order* both polynomials in *descending* powers of x, (ii) leave space, or insert zeros, for powers of x which have a zero coefficient.

For the sake of comparison the example below is related to Example 3.

Example 5 Divide $x^4 - x^3 - 3x^2 + 5x - 2$ by $x^2 + x - 2$.

$$
\begin{array}{r}
x^2 - 2x + 1 \\
x^2 + x - 2 \overline{\smash{\big)}\, x^4 - x^3 - 3x^2 + 5x - 2} \\
x^4 + x^3 - 2x^2 \\
\hline
-2x^3 - x^2 + 5x \\
-2x^3 - 2x^2 + 4x \\
\hline
x^2 + x - 2 \\
x^2 + x - 2 \\
\hline
0 + 0 + 0
\end{array}
$$

As we would expect, from Example 3, the division is exact, the *quotient* being $x^2 - 2x + 1$ and there is no *remainder*. Each term in the quotient is obtained by making the first term, x^2, in the *divisor*, $x^2 + x - 2$, divide exactly each time.

This working may, as in the case of multiplication, be written down, using detached coefficients:

$$\begin{array}{ccccccc}
 & & & x^4 & x^3 & x^2 & x & 1 \\
 & & & & & 1 & -2 & 1 \\
1 & 1 & -2\overline{)}1 & -1 & -3 & 5 & -2 \\
 & & 1 & 1 & -2 & & & \\
 & & \overline{} & -2 & -1 & 5 & & \\
 & & & -2 & -2 & 4 & & \\
 & & & \overline{} & 1 & 1 & -2 & \\
 & & & & 1 & 1 & -2 & \\
 & & & & \overline{} & . & . & . \\
\end{array}$$

Example 6 Divide $x^4 - 10$ by $x^2 + 3$.

In this example it is vital that we insert zeros for powers of x which have zero coefficients. The working is

$$\begin{array}{r}
x^2 + 0x - 3 \\
x^2 + 0x + 3\overline{)x^4 + 0x^3 + 0x^2 + 0x - 10} \\
x^4 + 0x^3 + 3x^2 \\
\overline{0x^3 - 3x^2 + 0x - 10} \\
-3x^2 + 0x - 9 \\
\overline{-1}
\end{array}$$

The quotient is $x^2 - 3$ and the remainder is -1. We can therefore write

$$x^4 - 10 \equiv (x^2 - 3)(x^2 + 3) - 1.$$

The remainder theorem

When the polynomial $P(x)$ is divided by the polynomial $\varphi(x)$, it is clear that the remainder $R(x)$ must have degree less than that of $\varphi(x)$. (If this is not so, the division is not complete.)

In particular, if $P(x)$ is of degree n and we divide it by $(x - \alpha)$, then the quotient $Q(x)$ will be a polynomial of degree $(n - 1)$ and the remainder $R(x)$ will be a constant. Thus,

$$P(x) \equiv (x - \alpha)Q(x) + R. \tag{2.3}$$

If we substitute $x = \alpha$, then we see that

$$P(\alpha) = R. \tag{2.4}$$

This result enables us to find the remainder without carrying out the division. It is known as the *remainder theorem*.

A slightly more general result is obtained by considering division by $(ax + b)$. Then

$$P_1(x) \equiv (ax + b)Q_1(x) + R_1. \qquad (2.5)$$
$$\Rightarrow P_1(-b/a) = R_1. \qquad (2.6)$$

(The value $(-b/a)$ is obtained by setting $ax + b = 0$.)

Example 7 Find the remainder when $P(x)$, where $P(x) \equiv 6x^3 - 7x^2 + 12x - 8$, is divided by (a) $(x - 1)$, (b) $(2x - 1)$.

(a) The remainder is $P(1) = 6 - 7 + 12 - 8 = 3$.

(b) The remainder is $P\left(\dfrac{1}{2}\right) = 6\left(\dfrac{1}{2}\right)^3 - 7\left(\dfrac{1}{2}\right)^2 + 12\left(\dfrac{1}{2}\right) - 8 = -3$.

The factor theorem

From Equation (2.6) we see that, if $P_1(-b/a) = 0$, then $R_1 = 0$
$\Rightarrow$ there is no remainder when $P_1(x)$ is divided by $ax + b$
$\Rightarrow$ $ax + b$ is a factor of $P_1(x)$.

This is known as the *factor theorem*.

From Equation (2.4) we see the result: 'If $P(\alpha) = 0$, then $(x - \alpha)$ is a factor of $P(x)$.'

Example 8 Show that $(x - 3)$ is a factor of $P(x)$, where $P(x) \equiv 2x^3 - 6x^2 + 9x - 27$.

$$P(3) = 2.27 - 6.9 + 9.3 - 27 = 0.$$

Therefore, $(x - 3)$ is a factor.

Example 9 Show that $(3x + 1)$ is a factor of $P(x)$, where $P(x) \equiv 6x^3 - x^2 - 19x - 6$.

$$3x + 1 = 0 \Rightarrow x = -\frac{1}{3}.$$

We consider, therefore, $P(-\tfrac{1}{3})$:

$$P\left(-\dfrac{1}{3}\right) = 6\left(-\dfrac{1}{3}\right)^3 - \left(-\dfrac{1}{3}\right)^2 - 19\left(-\dfrac{1}{3}\right) - 6$$
$$= -\dfrac{2}{9} - \dfrac{1}{9} + \dfrac{19}{3} - 6 = 0.$$

Therefore, $(3x + 1)$ is a factor.

Example 10 For what value of k is $(x + 3)$ a factor of $P(x)$, where $P(x) \equiv x^4 - 3x^2 + k$.

If $(x + 3)$ is a factor, $P(-3) = 0$.
$$P(-3) = 0 \Rightarrow 81 - 27 + k = 0$$
$$\Rightarrow k = -54.$$

Example 11 Given that $(x + 1)$ and $(x - 2)$ are factors of $P(x)$, where
$$P(x) \equiv ax^3 - bx^2 + ax + 6,$$
find the constants a and b.

Since $(x + 1)$ is a factor, $P(-1) = 0$,
$$\Rightarrow -a - b - a + 6 = 0 \Rightarrow -2a - b + 6 = 0.$$
Since $(x - 2)$ is a factor, $P(2) = 0$,
$$\Rightarrow 8a - 4b + 2a + 6 = 0 \Rightarrow 10a - 4b + 6 = 0.$$

Solving these simultaneous equations, we obtain $a = 1$ and $b = 4$.

Finding factors of polynomials

The factor theorem provides an important aid to finding factors of polynomials. To find linear factors, we seek values of λ such that $(x - \lambda)$ divides $P(x)$. As a consequence of the factor theorem, this is equivalent to finding values of λ for which $P(\lambda) = 0$.

If the polynomial is written in such a way that the coefficients are all integers, then it is suggested that one considers those values of λ which are a factor of the constant term a_0 in the polynomial.

Example 12 Find the factors of $P(x)$, where $P(x) \equiv x^3 - 4x^2 + x + 6$.

The only integer factors of 6 are $\pm 1, \pm 2, \pm 3, \pm 6$ and so these are the values of λ we consider.

Trial factor $(x - \lambda)$	$P(\lambda)$	Comment
(i) $(x - 1)$	$1 - 4 + 1 + 6 = 4$	Not zero, not a factor
(ii) $(x + 1)$	$-1 - 4 - 1 + 6 = 0$	$(x + 1)$ is a factor
(iii) $(x - 2)$	$8 - 16 + 2 + 6 = 0$	$(x - 2)$ is a factor
(iv) $(x + 2)$	$-8 - 16 - 2 + 6 = -10$	Not zero, not a factor
(v) $(x - 3)$	$27 - 36 + 3 + 6 = 0$	$(x - 3)$ is a factor
(vi) $(x + 3)$	$-27 - 36 - 3 + 6 = -60$	Not zero, not a factor

The factors are $(x + 1)$, $(x - 2)$ and $(x - 3)$ and the factorisation of $P(x)$ is
$$P(x) \equiv (x + 1)(x - 2)(x - 3).$$

At stage (ii), having found that $(x + 1)$ is a factor, we could, of course, divide $P(x)$ by $(x + 1)$ and deal with the remaining quadratic $x^2 - 5x + 6$, which immediately factorises into $(x - 2)(x - 3)$.

Also, at stage (iii) we have a factor $(x + 1)(x - 2)$ and the remaining factor can be obtained by dividing out or by inspection.

Stage (vi) is, of course, not necessary, as we have already found three factors of the cubic polynomial. For the same reason we need not consider $\lambda = \pm 6$.

The use of factors

There are several circumstances where it is advantageous to be able to write a polynomial $P(x)$ in terms of its factors. We will illustrate this by using the polynomial $P(x)$ considered in Example 12 above. We found that

$$P(x) \equiv x^3 - 4x^2 + x + 6 \equiv (x + 1)(x - 2)(x - 3).$$

Solution of equations

We recall the important property of number, 'if the product of two or more numbers is zero, then at least one of the numbers must be zero'. If we wish to solve the equation $P(x) = 0$, then, using the factors obtained above, we have

$$P(x) \equiv (x + 1)(x - 2)(x - 3) = 0.$$

By virtue of the above result, we have

$(x + 1) = 0$ or $(x - 2) = 0$ or $(x - 3) = 0$
$\Rightarrow x = -1$ or $x = 2$ or $x = 3$.

Curve sketching

If a curve has equation $y = P(x)$, then its intersections with the x-axis are given by $P(x) = 0$. Hence, from the above, the curve crosses the x-axis at $x = -1$, $x = 2$ and $x = 3$. [This is discussed in detail in *Curve Sketching*, by H.M. Kenwood and C. Plumpton (this series).]

In curve sketching it is also useful to know the stationary points, and these are given by $P'(x) = 0$. In our example

$$P'(x) = 3x^2 - 8x + 1.$$
$$P'(x) = 0 \Rightarrow 3x^2 - 8x + 1 = 0$$

$$\Rightarrow x = \frac{8 \pm \sqrt{(64 - 12)}}{6} = 2\cdot 54 \text{ or } 0\cdot 13 \text{ to two decimal places.}$$

[Stationary points are discussed in *Differentiation*, by C.T. Moss and C. Plumpton (this series).]

Solution of inequalities

The solution of the inequality $P(x) > 0$ is facilitated if we write down $P(x)$ in terms of its factors. Then we have

$$(x + 1)(x - 2)(x - 3) > 0.$$

The product of these numbers is only positive if either all the numbers are positive or two are negative and one positive. This situation is analysed in detail in Chapter 6 of this book.

2.2 Rational functions and partial fractions

A *rational algebraic function* is a function

$$f: x \mapsto \frac{P(x)}{Q(x)},$$

where $P(x)$ and $Q(x)$ are polynomials and $Q(x)$ is not the zero polynomial. Any values of x for which the denominator $Q(x)$ is zero must be excluded from the domain. We may write

$$f(x) = \frac{a_n x^n + a_{n-1} x^{n-1} + \cdots + a_1 x + a_0}{b_m x^m + b_{m-1} x^{m-1} + \cdots + b_1 x + b_0},$$

where m and n are integers and the a_is and b_js are constants. If the degree of $P(x)$ is less than the degree of $Q(x)$, i.e. $n < m$, then $f(x)$ is said to be of *proper* algebraic form.

To add together two rational expressions, we find the LCM of the denominators and express each fraction in an equivalent form with the LCM as denominator. For example,

$$f(x) \equiv \frac{5}{x+2} - \frac{4}{x+3}$$
$$\equiv \frac{5(x+3) - 4(x+2)}{(x+2)(x+3)} \equiv \frac{x+7}{(x+2)(x+3)}.$$

In several mathematical situations it is necessary to reverse this process and express complicated rational functions as the sum of simple proper algebraic fractions. This reverse process is called expressing $f(x)$ in terms of its *partial fractions*.

The procedure for expressing a rational algebraic function $f(x)$, where $f(x) = P(x)/Q(x)$, into partial fractions may be summarised as follows.

Step 1 If the degree of $P(x) \geq$ the degree of $Q(x)$, divide $P(x)$ by $Q(x)$ and obtain

$$P(x) = Q(x)S(x) + R(x).$$

Then

$$f(x) = S(x) + f_1(x), \quad \text{and} \quad f_1(x) = \frac{R(x)}{Q(x)}$$

is of proper algebraic form.

Step 2 We consider now $f(x)$, if it is of proper algebraic form, or $f_1(x)$ defined above.

Factorise the denominator into real linear and quadratic factors. (This is always possible.)

(i) For each linear factor $(ax + b)$, assume a partial fraction of the form

$$\frac{A}{(ax + b)},$$

where A is a constant.

(ii) For each repeated factor $(ax + b)^r$, assume there to be r *partial fractions* of the form

$$\frac{A_1}{(ax + b)}, \frac{A_2}{(ax + b)^2}, \ldots, \frac{A_r}{(ax + b)^r},$$

where $A_1, A_2, \ldots, A_r$ are constants.

(iii) For each quadratic factor $px^2 + qx + r$ assume there to be a partial fraction of the form

$$\frac{\alpha x + \beta}{px^2 + qx + r},$$

where α and β are constants. Here we do not discuss repeated quadratic factors.

Example 13 Express $\dfrac{x + 7}{(x + 2)(x + 3)}$ in partial fractions.

Assume from (i) that

$$f(x) \equiv \frac{x + 7}{(x + 2)(x + 3)} \equiv \frac{A}{(x + 2)} + \frac{B}{(x + 3)}$$
$$\Rightarrow (x + 7) \equiv A(x + 3) + B(x + 2). \qquad (2.7)$$

Equating coefficients of x,

$$A + B = 1,$$

Equating constant terms,

$$3A + 2B = 7.$$

Solving these simultaneous equations, we obtain $A = 5$ and $B = -4$. Therefore,

$$\frac{x + 7}{(x + 2)(x + 3)} \equiv \frac{5}{(x + 2)} - \frac{4}{(x + 3)}.$$

An alternative method for obtaining A and B is to use the fact that Equation (2.7) is an identity and holds for all values of x. If we substitute $x = -2$, we obtain

$$-2 + 7 = A(-2 + 3) \Rightarrow A = 5.$$

If we substitute $x = -3$, we obtain
$$-3 + 7 = B(-3 + 2) \Rightarrow B = -4.$$

This second method is often known as the 'cover-up rule', which may be stated quite generally:

If
$$\frac{g(x)}{(ax + b)(cx + d)} \equiv \frac{\gamma}{(ax + b)} + \frac{\delta}{(cx + d)},$$

where $g(x)$ is a linear function of x and γ and δ are constants, then
$$\frac{g(x)}{(cx + d)} = \gamma + \frac{\delta(ax + b)}{(cx + d)}.$$

Substituting $x = -b/a$ gives
$$\gamma = \frac{g(-b/a)}{(-cb/a + d)}.$$

This shows that the coefficient of $(ax + b)^{-1}$ in the partial fraction expansion can be obtained by 'covering up' the factor $ax + b$ in the original expression and putting $x = -b/a$ in the expression which remains.

[Strictly speaking, having multiplied by $(ax + b)$ we ought not to deduce any valid result by writing $x = -b/a$. However, it can be verified, by using a limiting procedure, that the results obtained in this way are in fact correct.]

One can of course use a combination of both methods and this will be illustrated in the example below.

Example 14 Express $\dfrac{2x^2 + 9x + 7}{(x^2 + 4)(2x - 3)}$ in partial fractions.

Using (i) and (iii) above, we write
$$\frac{2x^2 + 9x + 7}{(x^2 + 4)(2x - 3)} \equiv \frac{Ax + B}{(x^2 + 4)} + \frac{C}{(2x - 3)}$$
$$\Rightarrow 2x^2 + 9x + 7 \equiv (Ax + B)(2x - 3) + C(x^2 + 4). \qquad (2.8)$$

If we substitute $x = \tfrac{3}{2}$ into Equation (2.8), we obtain
$$\frac{9}{2} + \frac{27}{2} + 7 = C\left[\frac{9}{4} + 4\right]$$
$$\Rightarrow 25 = \frac{25}{4} C \Rightarrow C = 4.$$

Equating coefficients of $x^2 \Rightarrow 2 = 2A + C \Rightarrow A = -1$.
Equating constant terms $\Rightarrow 7 = -3B + 4C \Rightarrow B = 3$.
[It is a good idea to check that the values of the constants you have obtained satisfy Equation (2.8)]. Hence, the result is
$$\frac{(-x + 3)}{(x^2 + 4)} + \frac{4}{(2x - 3)}.$$

Example 15 Express $\dfrac{2x(5-x)}{(x-1)^2(x+3)}$ in partial fractions.

Using (i) and (ii) above, we write

$$\dfrac{2x(5-x)}{(x-1)^2(x+3)} \equiv \dfrac{A}{(x-1)^2} + \dfrac{B}{(x-1)} + \dfrac{C}{(x+3)}$$
$$\Rightarrow 10x - 2x^2 \equiv A(x+3) + B(x-1)(x+3) + C(x-1)^2. \quad (2.9)$$

Substitution of $x = 1$ gives

$$8 = 4A \Rightarrow A = 2.$$

Substitution of $x = -3$ gives

$$-48 = 16C \Rightarrow C = -3.$$

Equating coefficients of x^2 gives

$$-2 = B + C \Rightarrow B = 1.$$

Clearly, these values satisfy equation (2.9).
The required result is

$$\dfrac{2}{(x-1)^2} + \dfrac{1}{(x-1)} - \dfrac{3}{(x+3)}.$$

Example 16 Express $\dfrac{x^2 + 8x + 9}{(x+1)(x+2)}$ as the sum of a polynomial and partial fractions.

The given function is not a proper algebraic function but

$$x^2 + 8x + 9 \equiv 1(x^2 + 3x + 2) + (5x + 7).$$

Therefore,

$$\dfrac{x^2 + 8x + 9}{(x+1)(x+2)} = 1 + \dfrac{(5x+7)}{(x+1)(x+2)}. \quad (2.10)$$

We write the last term in Equation (2.10) in terms of partial fractions:

$$\dfrac{5x + 7}{(x+1)(x+2)} \equiv \dfrac{A}{(x+1)} + \dfrac{B}{(x+2)}.$$

Substituting $x = -1 \Rightarrow A = 2$.
Substituting $x = -2 \Rightarrow B = 3$.
Hence, the given function is equal to

$$1 + \dfrac{2}{(x+1)} + \dfrac{3}{(x+2)}.$$

Exercise 2

1. Add together $3x^3 + 2x + 6$, $2x^2 + x + 1$ and $2x^3 + x^2 + 3$.
2. Subtract $5x^3 + 2x^2 - 3x + 1$ from $6x^3 + 8x + 5$.
3. Multiply $(x^3 + 2x + 1)$ by $(x^2 + x + 1)$
   - (a) by multiplying the brackets,
   - (b) by long multiplication.
4. Divide $x^3 + 5x^2 + 11x + 10$ by $(x + 2)$.
5. Find the quotient and the remainder when $x^3 + 3x^2 + 2x$ is divided by $x^2 - x + 1$.
6. Find the remainder when
   - (a) $5x^3 + 2x^2 + x + 1$ is divided by $(x + 1)$,
   - (b) $6x^4 + 2x^3 + 3x^2 + x + 1$ is divided by $(2x - 1)$.
7. The polynomial $f(x) \equiv x^3 + ax^2 + bx + c$ leaves remainders 7, 1, 19 on division by $(x - 1)$, $(x + 1)$, $(x - 2)$, respectively. Find a, b, c and the remainder when $f(x)$ is divided by $(x + 2)$.
8. Show that $(x + 3)$ is a factor of $x^4 + 2x^3 + 7x^2 + 11x - 57$.
9. Find the values of a and b if $(x - 1)$ and $(x + 2)$ are both factors of $ax^3 + 3x^2 + bx - 2$, and state the third factor.
10. (a) Show that $(x - c)$ is a factor of $x^3 - c^3$ and hence show that
    $$x^3 - c^3 \equiv (x - c)(x^2 + cx + c^2).$$
    (b) Show that $(x + c)$ is a factor of $x^3 + c^3$ and hence show that
    $$x^3 + c^3 \equiv (x + c)(x^2 - cx + c^2).$$
11. By using the factor theorem, find one factor of $x^3 - x^2 + 2x - 8$ and hence factorise the expression.
12. Find the factors of $3x^3 + 5x^2 - 16x - 12$.
13. Express as a single rational function
    - (a) $\dfrac{2}{(x - 2)} + \dfrac{3}{(x + 3)}$,
    - (b) $\dfrac{2}{(x^2 + x + 2)} - \dfrac{3}{(x + 3)}$.
14. Express in partial fractions
    - (a) $\dfrac{2x + 2}{(x - 1)(x + 3)}$,
    - (b) $\dfrac{3x^2 - 2}{(x + 1)(x^2 + x + 1)}$,
    - (c) $\dfrac{7x + 4}{(x - 3)(x + 2)^2}$.

    Check all your answers by recombining the partial fractions.
15. Express $\dfrac{x^3 - 4x + 5}{(x - 2)(x + 3)}$ as the sum of a polynomial and partial fractions.

3 The quadratic function and quadratic equations

3.1 Quadratic functions

The function f(x), where $f(x) \equiv ax^2 + bx + c$, and a, b, c are constants, $a \neq 0$, is called a *quadratic function*, or sometimes a quadratic polynomial. From elementary algebra

$$(x + d)^2 \equiv x^2 + 2dx + d^2.$$

Using this, we write

$$ax^2 + bx + c \equiv a\left(x^2 + \frac{b}{a}x + \frac{c}{a}\right) \equiv a\left[\left(x + \frac{b}{2a}\right)^2 + \frac{4ac - b^2}{4a^2}\right]$$

$$\Rightarrow ax^2 + bx + c \equiv a\left(x + \frac{b}{2a}\right)^2 + \frac{4ac - b^2}{4a}. \tag{3.1}$$

The quadratic function f(x) has the following properties:
(i) $f(0) = c$.
(ii) As $x \to \pm\infty$; if $a > 0$, $f(x) \to +\infty$,
if $a < 0$, $f(x) \to -\infty$.
(iii) If $a > 0$, then, from Equation (3.1),

$$f(x) \geq \frac{4ac - b^2}{4a}.$$

The minimum value $(4ac - b^2)/(4a)$ is attained when $x = -b/(2a)$.
If $a < 0$, then

$$f(x) \leq \frac{4ac - b^2}{4a}.$$

The maximum value $(4ac - b^2)/(4a)$ is attained when $x = -b/(2a)$.
(iv) $f(x) = 0 \Rightarrow a\left[\left(x + \frac{b}{2a}\right)^2 + \frac{4ac - b^2}{4a^2}\right] = 0$

$$\Rightarrow \left(x + \frac{b}{2a}\right)^2 = \frac{b^2 - 4ac}{4a^2}$$

$$\Rightarrow \left(x + \frac{b}{2a}\right) = \pm\frac{\sqrt{b^2 - 4ac}}{2a}$$

$$\Rightarrow x = \frac{-b \pm \sqrt{b^2 - 4ac}}{2a}. \tag{3.2}$$

This result has an important graphical interpretation. There are three cases to consider, depending on whether the *discriminant* $\Delta = b^2 - 4ac$ is positive, negative or zero.

(1) If $b^2 > 4ac$, the curve $y = f(x)$ cuts the x-axis at two distinct points. This, taken together with (ii), enables us to draw the sketches of the curves $y = f(x)$ as in Fig. 3.1.

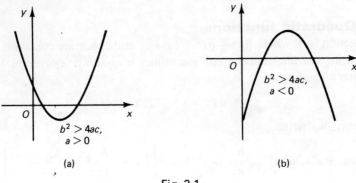

Fig. 3.1

(2) If $b^2 = 4ac$, the curve $y = f(x)$ touches the x-axis, since the equation $f(x) = 0$ then has two equal roots at $x = -b/(2a)$.

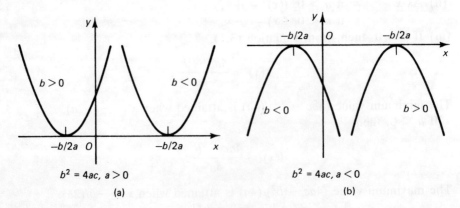

Fig. 3.2

(3) If $b^2 < 4ac$, the equation $f(x) = 0$ has no real roots.
[We have seen above that when $\Delta < 0$, there are no real roots of the equation $f(x) = 0$. In such a situation there are always two complex roots. If we define i to be such that $i^2 = -1$, then, from formula (3.2),

$$x = \frac{-b \pm \sqrt{(-1)(4ac - b^2)}}{2a} = \frac{-b}{2a} \pm i\frac{\sqrt{(4ac - b^2)}}{2a}.$$
]

30 *Methods of Algebra*

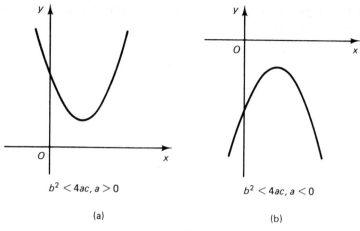

$b^2 < 4ac, a > 0$ (a)

$b^2 < 4ac, a < 0$ (b)

Fig. 3.3

3.2 Quadratic equations

The equation $ax^2 + bx + c = 0$, where $a \neq 0$, is called a *quadratic equation*. From Equation (3.2) the roots of this quadratic equation are

$$\frac{-b \pm \sqrt{(b^2 - 4ac)}}{2a}. \qquad (3.3)$$

If we denote these roots by α and β, then

$$ax^2 + bx + c = a(x - \alpha)(x - \beta). \qquad (3.4)$$

From Equation (3.3), or alternatively by equating coefficients of x^1 and x^0 in Equation (3.4), we find

$$\alpha + \beta = -b/a, \qquad (3.5a)$$
$$\alpha\beta = c/a. \qquad (3.5b)$$

It follows immediately from Equation (3.5b) that, when the roots are real, they are of the *same sign* if a and c are of the *same sign*. (From Equation (3.5a) it follows that they are of different sign from b.) The roots are of *opposite sign* if a and c are of *opposite sign*.

The above analysis indicates the following:

$$\left\{ f(x) = 0 \text{ has } \begin{matrix} \text{two distinct} \\ \text{two equal} \\ \text{no} \end{matrix} \text{ real roots} \right\} \Leftrightarrow \left\{ b^2 - 4ac \begin{matrix} > 0 \\ = 0 \\ < 0 \end{matrix} \right\}.$$

To solve a quadratic equation it is often not necessary to use the general formula (3.3). In simple cases, when the roots are integers or rational numbers, one can often express the quadratic function in terms of its factors, as in Equation (3.4), by inspection. The roots are then just α and β. If a factorisation is not immediately obvious, one may try the factor theorem (p. 21) to obtain a factor.

Example 1 Describe the important features and sketch the graph of:
(a) $f_1(x) = x^2 - 3x + 5$,
(b) $f_2(x) = x^2 - 5x + 6$,
(c) $f_3(x) = -x^2 + 2x - 1$.

(a) $f_1(x) = x^2 - 3x + 5$.
  (i) $f_1(0) = 5$.
  (ii) As $x \to \pm\infty$, $f_1(x) \to \infty$.
  (iii) $f_1(x) = (x^2 - 3x + 5)$
$$= \left[x^2 - 3x + \left(\frac{3}{2}\right)^2\right] - \left(\frac{3}{2}\right)^2 + 5$$
$$= \left(x - \frac{3}{2}\right)^2 + \frac{11}{4} \geq \frac{11}{4}.$$

So $f_1(x)$ has a minimum value of $\dfrac{11}{4}$ when $x = \dfrac{3}{2}$, and so is never zero. The graph of $f_1(x)$ is shown in Fig. 3.4.

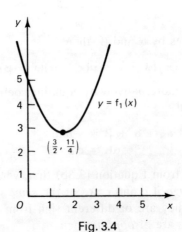

Fig. 3.4

(b) $f_2(x) = x^2 - 5x + 6$.
  (i) $f_2(0) = 6$.
  (ii) As $x \to \pm\infty$, $f_2(x) \to +\infty$.
  (iii) $f_2(x) = x^2 - 5x + 6 \equiv \left[x^2 - 5x + \left(\frac{5}{2}\right)^2\right] - \left(\frac{5}{2}\right)^2 + 6$
$$\equiv \left(x - \frac{5}{2}\right)^2 - \frac{1}{4}.$$

So $f_2(x)$ has a minimum value of $-\frac{1}{4}$ when $x = \frac{5}{2}$.

(iv) $f_2(x) = 0 \Rightarrow \left(x - \frac{5}{2}\right)^2 - \frac{1}{4} = 0$

$\Rightarrow \left(x - \frac{5}{2}\right)^2 = \frac{1}{4}$

$\Rightarrow \left(x - \frac{5}{2}\right) = \pm\frac{1}{2}$

$\Rightarrow x = 3$ or 2.

Alternatively we could write

$$f_2(x) = (x - 3)(x - 2)$$
$$\Rightarrow f_2(x) = 0 \text{ when } x = 3 \text{ or } 2.$$

The graph of $f_2(x)$ is shown in Fig. 3.5.

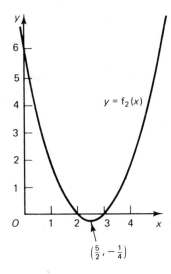

Fig. 3.5

(c) $f_3(x) = -x^2 + 2x - 1$.
 (i) $f_3(0) = -1$.
 (ii) As $x \to \pm\infty$, $f_3(x) \to -\infty$.
 (iii) $f_3(x) = -x^2 + 2x - 1$
$\equiv -(x^2 - 2x) - 1$
$\equiv -(x^2 - 2x + 1) + 1 - 1$
$\equiv -(x - 1)^2$.

So $f_3(x)$ has a maximum value of 0 when $x = 1$.

(iv) $f_3(x) = 0 \Rightarrow (x-1)^2 = 0$
$\Rightarrow x = 1$ twice.
The graph of $f_3(x)$ is shown in Fig. 3.6.

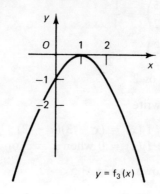

Fig. 3.6

Example 2 Defining $g_1(x) = 1/[f_1(x)]$, $g_2(x) = 1/[f_2(x)]$ and $g_3(x) = 1/[f_3(x)]$, where $f_1(x)$, $f_2(x)$ and $f_3(x)$ are given in Example 1, describe the important features of $g_1(x)$, $g_2(x)$ and $g_3(x)$ and sketch the graph of each of them.

(a) $g_1(x) = \dfrac{1}{f_1(x)} = \dfrac{1}{x^2 - 3x + 5}$.

(i) $g_1(0) = \frac{1}{5}$.
(ii) As $x \to \pm\infty$, $g_1(x) \to 0$ from above.
(iii) Since the minimum value of $f_1(x)$ is $\frac{11}{4}$ and is attained at $x = \frac{3}{2}$, the maximum value of $g_1(x)$ is $\frac{4}{11}$ when $x = \frac{3}{2}$. Further, since $f_1(x)$ never vanishes, $g_1(x)$ always remains finite. The graph of $g_1(x)$ is shown in Fig. 3.7.

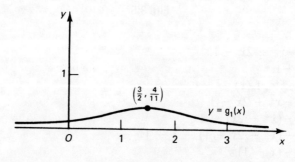

Fig. 3.7

(b) $g_2(x) = \dfrac{1}{f_2(x)} = \dfrac{1}{x^2 - 5x + 6}$.

(i) $g_2(0) = \frac{1}{6}$.

(ii) As $x \to \pm\infty$, $g_2(x) \to 0$ from above.

(iii) Since $f_2(x)$ has a local minimum value of $-\frac{1}{4}$ at $x = \frac{5}{2}$, $g_2(x)$ has a local maximum value of -4 at $x = \frac{5}{2}$.

(iv) Since $f_2(x) = 0$ when $x = 2$, $x = 3$, the graph of $g_2(x)$ has asymptotes $x = 2$ and $x = 3$.

Further,

$$f_2(x) > 0 \text{ when } x < 2 \text{ or } x > 3$$

$\Rightarrow g_2(x) > 0$ when $x < 2$ or $x > 3$,
and

$$f_2(x) < 0 \text{ when } 2 < x < 3$$

$\Rightarrow g_2(x) < 0$ when $2 < x < 3$.

The graph of $g_2(x)$ is shown in Fig. 3.8.

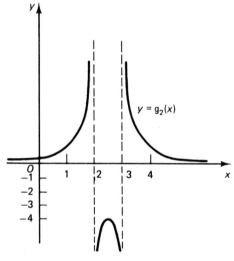

Fig. 3.8

(c) $g_3(x) = \dfrac{1}{f_3(x)} = \dfrac{1}{-x^2 + 2x - 1}$.

(i) $g_3(0) = -1$.

(ii) As $x \to \pm\infty$, $g_3(x) \to 0$ from below.

(iii) Since $f_3(x) = 0$ when $x = 1$, $g_3(x)$ has an asymptote at $x = 1$.

Further, as $f_3(x) < 0$ ($x \neq 1$),

$$g_3(x) < 0 \text{ for all } x \neq 1.$$

The graph of $g_3(x)$ is shown in Fig. 3.9.

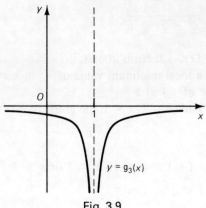

Fig. 3.9

Example 3 Find the set of values for which
(a) $x^2 > 5x - 6$,
(b) $x^2 + 5 < 4x$,
(c) $3x^2 + x < 2$.

(a) Let $f(x) \equiv x^2 - 5x + 6$.
$$x^2 > 5x - 6 \Leftrightarrow f(x) > 0.$$

We have already sketched the graph of $f(x)$ in Fig. 3.5, from which it is clear that $f(x) > 0$ if $x < 2$ or $x > 3$. The required set is, therefore,
$$\{x\colon x < 2\} \cup \{x\colon x > 3\}.$$

(b) Let $g(x) \equiv x^2 - 4x + 5$.
$$x^2 + 5 < 4x \Leftrightarrow g(x) < 0.$$
However,
$$g(x) \equiv (x - 2)^2 + 1,$$
so that $g(x)$ has a minimum value of 1 at $x = 2$. Hence, $g(x)$ is never less than 0 and there are no real values of x for which $x^2 + 5 < 4x$.

(c) Let $h(x) \equiv 3x^2 + x - 2$.
$$3x^2 + x < 2 \Leftrightarrow h(x) < 0.$$
$$h(x) \equiv 3\left[x^2 + \frac{x}{3} - \frac{2}{3}\right]$$
$$= 3\left[\left(x + \frac{1}{6}\right)^2 - \frac{1}{36} - \frac{2}{3}\right] = 3\left[\left(x + \frac{1}{6}\right)^2 - \frac{25}{36}\right].$$
$$h(x) = 0 \Rightarrow x + \frac{1}{6} = \pm\frac{5}{6}$$

$\Rightarrow x = -1$ or $\frac{2}{3}$.

In addition, h(x) → ∞ as x → ±∞ and h(0) = −2.
The graph of h(x) is shown in Fig. 3.10. From the graph we see that h(x) < 0 for $-1 < x < \frac{2}{3}$, i.e. the required set is

$$\{x: -1 < x < \tfrac{2}{3}\}.$$

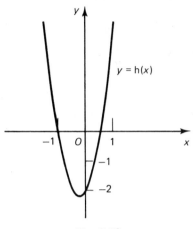

Fig. 3.10

Example 4 Solve, for real values of x, the quadratic equations
(a) $3x^2 + 4x - 3 = 0$,
(b) $4x^2 - 28x + 49 = 0$,
(c) $x^2 - x + 1 = 0$.

(a) $3x^2 + 4x - 3 = 0$. Here $a = 3$, $b = 4$, $c = -2$. Using the formula we obtain

$$x = \frac{-4 \pm \sqrt{[4^2 - 4 \times 3 \times (-3)]}}{6}$$

$$= \frac{-4 \pm \sqrt{52}}{6} \approx \frac{-4 \pm 7\cdot 211}{6} \quad \text{(using a calculator to find } \sqrt{52}\text{)}$$

$$\approx \frac{3\cdot 211}{6} \text{ or } \frac{-11\cdot 211}{6}$$

$\Rightarrow x \approx 0\cdot 535$ or $-1\cdot 869$.

(b) $4x^2 - 28x + 49 = 0$. Here $a = 4$, $b = -28$, $c = 49$.
Hence,

$$x = \frac{28 \pm \sqrt{[(28)^2 - 4 \times 4 \times 49]}}{8} = \frac{28 \pm 0}{8}$$

$\Rightarrow x = 3\tfrac{1}{2}$ twice.

Alternatively we might notice that $4x^2 - 28x + 49 \equiv (2x - 7)^2$, which gives the same answer immediately.
(c) $x^2 - x + 1 = 0$. Here $a = 1$, $b = -1$, $c = 1$.
Hence,
$$x = \frac{1 \pm \sqrt{(1 - 4)}}{2} = \frac{1 \pm \sqrt{(-3)}}{2}$$
$\Rightarrow x^2 - x + 1 = 0$ has no real solutions.

Example 5 If α and β are the roots of the equation $ax^2 + bx + c = 0$, find the values of (a) $\alpha^2 + \beta^2$, (b) $\dfrac{1}{\alpha} + \dfrac{1}{\beta}$, (c) $\alpha^3 + \beta^3$.

(a) We recall that $\alpha + \beta = -b/a$, $\alpha\beta = c/a$. We therefore seek to write $\alpha^2 + \beta^2$ in terms of $(\alpha + \beta)$ and $(\alpha\beta)$. In fact
$$\alpha^2 + \beta^2 \equiv [(\alpha + \beta)^2 - 2\alpha\beta]$$
$$= \frac{b^2}{a^2} - 2\frac{c}{a} = \frac{b^2 - 2ac}{a^2}$$

using the above.

(b) $\dfrac{1}{\alpha} + \dfrac{1}{\beta} \equiv \dfrac{\beta + \alpha}{\alpha\beta} = \left(\dfrac{-b}{a}\right) \Big/ \left(\dfrac{c}{a}\right) = \dfrac{-b}{c}$.

(c) $\alpha^3 + \beta^3 \equiv (\alpha + \beta)(\alpha^2 - \alpha\beta + \beta^2)$
$\equiv (\alpha + \beta)[(\alpha + \beta)^2 - 3\alpha\beta]$
$= \left(\dfrac{-b}{a}\right)\left[\dfrac{b^2}{a^2} - \dfrac{3c}{a}\right]$
$= b(3ac - b^2)/a^3$.

Example 6 If α and β are the roots of the equation
$$2x^2 + 3x + 1 = 0,$$
form an equation for which the roots are α/β and β/α.

The required equation is
$$x^2 - \text{(the sum of the roots)} \, x + \text{(the product of the roots)} = 0,$$
i.e.
$$x^2 - \left(\frac{\alpha}{\beta} + \frac{\beta}{\alpha}\right)x + 1 = 0.$$

But
$$\frac{\alpha}{\beta} + \frac{\beta}{\alpha} \equiv \frac{\alpha^2 + \beta^2}{\alpha\beta}.$$

From the given equation $\alpha + \beta = -\frac{3}{2}$ and $\alpha\beta = \frac{1}{2}$
$\Rightarrow \alpha^2 + \beta^2 \equiv (\alpha + \beta)^2 - 2\alpha\beta = \frac{9}{4} - 1 = \frac{5}{4}$.
Hence, the required equation is

$$x^2 - \left(\frac{5}{4}\right)\bigg/\left(\frac{1}{2}\right)x + 1 = 0$$
$$\Leftrightarrow x^2 - \tfrac{5}{2}x + 1 = 0 \Leftrightarrow 2x^2 - 5x + 2 = 0.$$

Example 7 A stone thrown vertically upwards with speed 20 m s^{-1} is at a height y m above the point of projection O after t s, where

$$y = 20t - 5t^2.$$

For how long is the stone more than 10 m above O.

We require

$$20t - 5t^2 > 10$$
or
$$-5t^2 + 20t - 10 > 0.$$

Let $f(t) = -5t^2 + 20t - 10$. Then
(i) $f(0) = -10$.
(ii) As $t \to \pm\infty$, $f(t) \to -\infty$.

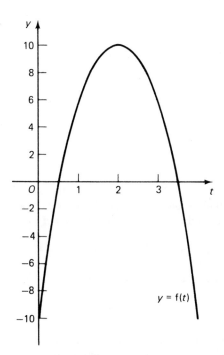

Fig. 3.11

(iii) $f(t) = 0 \Rightarrow -5t^2 + 20t - 10 = 0$
$$\Rightarrow t^2 - 4t + 2 = 0$$
$$\Rightarrow t = \frac{4 \pm \sqrt{(16-8)}}{2} = \frac{4 \pm \sqrt{8}}{2}.$$

Let us call these roots t_1 and t_2. Then a sketch of $f(t)$ is given in Fig. 3.11. It is clear from the sketch that $f(t) > 0$ for $t_1 < t < t_2$, i.e. for a length of time $(t_2 - t_1)$ s $= \sqrt{8}$ s. Hence, the stone is more than 10 m above O for $\sqrt{8}$ s $\approx 2 \cdot 83$ s.

Exercise 3

1. Describe the important features and sketch the graph of
   (a) $f(x) = 4x^2 + 4x + 1$,
   (b) $g(x) = 4 - 3x - x^2$,
   (c) $h(x) = x^2 - x + 3$.
   Sketch also $\dfrac{1}{f(x)}$, $\dfrac{1}{g(x)}$ and $\dfrac{1}{h(x)}$.

2. Solve the quadratic equations
   (a) $3x^2 - 2x + 1 = 0$,
   (b) $x^2 + 3x - 2 = 0$,
   (c) $4x^2 - 20x + 24 = 0$,
   (d) $9x^2 - 6x + 1 = 0$.

3. Find the ranges of values of x for which
   (a) $x^2 > x$,
   (b) $x^2 < x - 2$,
   (c) $x^2 \leqslant 1 - x$,
   (d) $x^2 > x - 1$.

4. If α and β are the roots of the equation
   $$2x^2 - 4x + 1 = 0,$$
   without solving this equation, form equations whose roots are
   (a) $\dfrac{1}{\alpha}, \dfrac{1}{\beta}$,
   (b) $(2\alpha + \beta), (\alpha + 2\beta)$,
   (c) $\dfrac{1}{\alpha^2}, \dfrac{1}{\beta^2}$.

5. Given that $f(x) = px^2 - 2x + 3p + 2$, find the two values of p for which the equation $f(x) = 0$ has equal roots. Find also the set of values of p for which $f(x)$ is negative for all real values of x. Sketch the graph of $y = f(x)$ for each of the cases $p = -2$, $p = 1$.

6. Given that one of the roots of the equation $x^2 + 3x + c = 0$ is twice the other, find the value of c.

7. A batsman hits a cricket ball so that its trajectory has the equation
   $$y = x - \frac{1}{60}x^2,$$
   where x and y are horizontal and vertical distances, measured in metres. How far away does it land on horizontal ground and what is its greatest height above the ground?

8 The roots of the equation
$$9x^2 + 6x + 1 = 4kx,$$
where k is a real constant, are denoted by α and β.
(a) Show that an equation whose roots are $1/\alpha$ and $1/\beta$ is
$$x^2 + 6x + 9 = 4kx.$$
(b) Find the set of values of k for which α and β are real.
(c) Find also the set of values of k for which α and β are real and positive.

4 Mathematical proof

4.1 Some logical concepts

Mathematics deals with numbers and symbols, but what really distinguishes mathematics from other sciences is the use of proof. A scientific theory may be supported by several thousand observations but can never be proved by observation. There is always the possibility that some observer will produce some contradictory evidence.

In mathematics we deal with statements or propositions. For our present purpose, we define a *statement* as a sentence which is either true or false, but not both. A proof consists of a sequence of logical steps leading from a set of known statements to the new statement which is being proved.

Each of the logical steps by means of which an argument advances is of the form 'if statement P is true, then it follows that statement Q is true'. This is usually abbreviated to 'if P then Q' or 'P implies Q'. In symbols we write $P \Rightarrow Q$. The soundness of the step does not depend on whether P is a true fact. For example, the argument

$$(\text{one egg costs £5}) \Rightarrow (\text{four eggs cost £20})$$

is valid whatever the actual price of eggs.

An alternative way of writing $P \Rightarrow Q$ is $Q \Leftarrow P$, which means 'Q is implied by P'.

Example 1 (P is the mid-point of the straight line segment AB) $\Rightarrow (PA = PB)$.

The statement $Q \Rightarrow P$ is the *converse* of the statement $P \Rightarrow Q$. If $P \Rightarrow Q$ is a valid statement, its converse may or may not be valid also.

Example 2

$$(x = 4) \Rightarrow (x^2 = 16).$$

But

$$(x^2 = 16) \Rightarrow (\text{either } x = 4 \text{ or } x = -4).$$

A common error of reasoning is to establish the converse instead of what is required to be proved.

Example 3 Prove that $y = mx + c$ is a tangent to $x^2 + y^2 = a^2$ if
$$c^2 = a^2(1 + m^2).$$
The following is often presented. If $y = mx + c$ is a tangent to $x^2 + y^2 = a^2$, then

$x^2 + (mx + c)^2 = a^2$ and the equation has equal roots
$\Rightarrow [(1 + m^2)x^2 + 2mcx + c^2 - a^2 = 0$ has equal roots]
$\Rightarrow [4m^2c^2 = 4(1 + m^2)(c^2 - a^2)]$
$\Rightarrow [c^2 = a^2(1 + m^2)].$

This is the converse of the result to be proved. The correct proof is as follows:
[equation $c^2 = a^2(1 + m^2)$] $\Rightarrow$ [equation in x has equal roots]
$\Rightarrow$ (line $y = mx + c$ has double contact with $x^2 + y^2 = a^2$)
$\Rightarrow$ (line is a tangent).
Care should be taken in this regard.

If $P \Rightarrow Q$, we say that P is a *sufficient* condition for Q. If $P \Leftarrow Q$, we say that P is a *necessary* condition for Q. When $P \Rightarrow Q$ and $Q \Rightarrow P$, we say that P is *equivalent* to Q and write $P \Leftrightarrow Q$. This means 'P if and only if Q' and we say that P is a *necessary and sufficient* condition for Q.

That implication and equivalence are not the same is shown by the following example.

Example 4 Solve the equation $\sqrt{(3x)} - \sqrt{(x + 1)} = 1$. We write the equation as $\sqrt{(3x)} - 1 = \sqrt{(x + 1)}$. Squaring both sides, we obtain

$3x + 1 - 2\sqrt{(3x)} = x + 1$
$\Rightarrow 2x = 2\sqrt{(3x)}$ or $x = \sqrt{(3x)}$
$\Rightarrow x^2 = 3x$ (squaring both sides)
$\Rightarrow x = 3$ or $x = 0.$

Hence,
$$\sqrt{(3x)} - \sqrt{(x + 1)} = 1 \Rightarrow x = 0, 3.$$

This is not necessarily reversible.

$x = 3 \Rightarrow \sqrt{(3x)} - \sqrt{(x + 1)} = 1.$ So $x = 3$ is a solution.
$x = 0 \Rightarrow \sqrt{(3x)} - \sqrt{(x + 1)} = -1.$ So $x = 0$ is not a solution.

[The reason for the occurrence of the false root $x = 0$ is that in the above we have squared both sides of the equation on two occasions and included the root of $\sqrt{(3x)} - \sqrt{(x + 1)} = -1$ also.]

The *negation* of a statement P is the statement P' (or $\sim P$) which is false when P is true and true when P is false. P' (or $\sim P$) is often read 'not P'.

Example 5
(a) If P is the statement $(x = 2)$, then P' (or $\sim P$) is the statement $(x \neq 2)$.
(b) If P is the statement (C lies on AB) then P' (or $\sim P$) is the statement (C does not lie on AB).

Example 6 Suppose P is the statement $(x = 2)$,
Q is the statement $(x^2 = 4)$.
Then
$$P': (x \neq 2),$$
$$Q': (x^2 \neq 4).$$

Notice that
$$P \Rightarrow Q \quad \text{(i.e. if } x = 2 \text{ then } x^2 = 4\text{)},$$
$$Q' \Rightarrow P' \quad \text{(i.e. if } x^2 \neq 4 \text{ then } x \neq 2\text{)}.$$

Note that the statements $Q \Rightarrow P$ and $P' \Rightarrow Q'$ are both false. Each of the two statements $P \Rightarrow Q$ and $Q' \Rightarrow P'$ is called the *contrapositive* of the other. We have the following relationship: 'a statement and its contrapositive are either both true or both false', so that

$$P \Rightarrow Q \text{ and } Q' \Rightarrow P' \text{ are equivalent statements.}$$

4.2 Proof by contradiction (*reductio ad absurdum*)

By negating the negation we return to the original statement, so that $(P')'$ is the same as P. This result is the basis of a powerful method of proof called 'proof by contradiction'. We can prove the truth of P by showing that P' is false.

Example 7 Prove that there are infinitely many prime numbers. (A prime number has no factors other than 1 and itself.)

We assume the negated statement

(the number of primes is finite)
$\Rightarrow$ (there exists an integer p, such that p is the largest prime).

Consider the number $p! + 1$. This is not divisible by p or by any positive integer less than p (the remainder is 1 in every case)
$\Rightarrow$ either $(p! + 1)$ is not divisible by an integer other than 1 or $(p! + 1)$, in which case $p! + 1$ is a prime, or $p! + 1$ is divisible by a number between p and $(p! + 1)$
$\Rightarrow$ there is a prime number larger than p.
The assumption 'the number of primes is finite'
$\Rightarrow$ (i) p is the largest prime,
  (ii) there is a prime larger than p.

(i) and (ii) are contradictory. Hence, the statement (the number of primes is finite) is false and so the statement (the number of primes is infinite) is true.

4.3 The use of a counter-example

To prove that a statement is false is frequently much easier than proving that the statement is true. All that is necessary, in order to prove that a statement is false, is to produce *just one case* for which the statement is in fact false. This (contradictory) case is called a *counter-example*.

Example 8 Find counter-examples to show that the following statements are false:
(a) $(x^2 = y^2) \Rightarrow (x = y)$,
(b) $(a - b > 0) \Rightarrow (a^2 - b^2 > 0)$,
(c) (all odd numbers are prime).

(a) $x = 3$, $y = -3$ satisfy $x^2 = y^2$ but not $x = y$. Statement is false.
(b) If $a = 1$, $b = -2$, then $a - b = 3$, which is >0 but $a^2 - b^2 = -3$, which is <0.
(c) 9 is an odd number but it is not prime.

4.4 Proof by deduction

Proof by deduction is a method of proof which we have already used in this book. It is the method of proof which is probably most familiar to the reader and is often used without the word 'proof' being mentioned.

Essentially, to prove $P \Rightarrow Q$ we proceed by several intermediate stages, proving $P \Rightarrow R$, then $R \Rightarrow S$ and then $S \Rightarrow Q$. (There may, of course, be more than two intermediate stages.)

Example 9 Prove that
$$(x^2 - 5x + 6 = 0) \Rightarrow (x = 2, 3).$$
The statement $(x^2 - 5x + 6 = 0)$
$$\Leftrightarrow [(x - 3)(x - 2) = 0]$$
$$\Leftrightarrow [x - 3 = 0 \text{ or } x - 2 = 0]$$
$$\Leftrightarrow [x = 3 \text{ or } x = 2].$$
Here we have proved both the result and its converse.

4.5 Proof by exhaustion

The method of proof by exhaustion (not a state of mind!) is of limited use and is restricted to situations in which there are only a finite number of possibilities which may each be examined in turn.

Example 10 Prove that there is no solution, in integers, of the equation $x^2 + y^2 = 11$.

If $x = 1$, then

$y = 1$ gives $x^2 + y^2 = 2$,
$y = 2$ gives $x^2 + y^2 = 5$,
$y = 3$ gives $x^2 + y^2 = 10$,
$y = 4$ gives $x^2 + y^2 > 11$.

If $x = 2$, then

$y = 1$ gives $x^2 + y^2 = 5$,
$y = 2$ gives $x^2 + y^2 = 8$,
$y = 3$ gives $x^2 + y^2 > 11$.

If $x = 3$, then

$y = 1$ gives $x^2 + y^2 = 10$,
$y = 2$ gives $x^2 + y^2 > 11$.

If $x = 4$, then

$$x^2 + y^2 > 11.$$

Thus, we have exhausted all possibilities. It is clear that we need not consider negative values. It is probably also clear that we could have reduced the labour by using the fact that the given equation is symmetrical in x and y and so we need only consider solutions in which $y \geq x$.

4.6 Proof by mathematical induction

Proof by mathematical induction is a very general method of proof once a possible result has been suspected or conjectured. It is not usually a means of discovering new results but a method of proving (formally) results expected to be true.

This method of proof applies to statements (S) which involve a positive integer n. We wish to prove that the statement (S) is true for all integers n greater than some fixed integer n_0. The fixed integer n_0 is usually 1 but this need not be so. (See Examples 14 and 15.)

The proof has two distinct steps.
(i) Show that the statement (S) is true for the value n_0 of n.
(ii) Show that, if the statement (S) is assumed to be true for a particular value of n, $n = k$ say, then (S) is true for the next value of n — that is, for $n = k + 1$.

Using (ii), we can establish the truth of the statement for each successive value of n starting from the value in (i).

This method of proof may be compared with the process of climbing upstairs. If we can (i) reach a starting place somewhere on the stairs and (ii) get from one stair to the next, then we can climb as far as we wish up the stairs.

Example 11 Prove by induction that, for $n \in \mathbb{Z}^+$,

$$1 + 2 + 3 + \cdots + n = \tfrac{1}{2}n(n + 1).$$

When $n = 1$,

$$\text{LHS (left-hand side)} = 1$$
$$\text{RHS (right-hand side)} = (\tfrac{1}{2}.2) = 1.$$

Hence, the statement is true for $n = 1$.
Assume the statement is true for $n = k$, that is,

$$1 + 2 + 3 + \cdots + k = \tfrac{1}{2}k(k + 1). \tag{4.1}$$

Then the sum of $(k + 1)$ terms on the LHS is

$$\begin{aligned} 1 + 2 + 3 + \cdots + k + (k + 1) &= \tfrac{1}{2}k(k + 1) + (k + 1), \text{ using Equation (4.1)} \\ &= \tfrac{1}{2}k(k + 1)(k + 2), \\ &= \tfrac{1}{2}(k + 1)[(k + 1) + 1]. \end{aligned}$$

This is just the RHS of Equation (4.1) with k replaced by $(k + 1)$. Hence, if the statement is true for $n = k$, it is true for $n = k + 1$. But the statement is true for $n = 1$. Therefore it is true for $n = 1 + 1 = 2$. Similarly, it is true for $n = 2 + 1 = 3$, and so on.

Therefore, by induction, the statement is true for all integers $n \geq 1$ or $n \in \mathbb{Z}^+$.

Example 12 Prove that, for $n \in \mathbb{Z}^+$,

$$1^3 + 2^3 + 3^3 + \cdots + n^3 = [\tfrac{1}{2}n(n + 1)]^2.$$

When $n = 1$,

$$\text{LHS} = 1,$$
$$\text{RHS} = [\tfrac{1}{2}.2]^2 = 1.$$

Hence, the statement is true for $n = 1$.
Assume the statement is true for $n = k$, that is,

$$1^3 + 2^3 + 3^3 + \cdots + k^3 = [\tfrac{1}{2}k(k + 1)]^2 \tag{4.2}$$

When $n = k + 1$, the LHS is

$$\begin{aligned} 1^3 + 2^3 + 3^3 + \cdots + k^3 + (k + 1)^3 &= [\tfrac{1}{2}k(k + 1)]^2 + (k + 1)^3 \text{ by Equation (4.2)} \\ &= \tfrac{1}{4}(k + 1)^2[k^2 + 4(k + 1)] \\ &= \tfrac{1}{4}(k + 1)^2[k^2 + 4k + 4] \end{aligned}$$

$$= \tfrac{1}{4}(k+1)^2(k+2)^2$$
$$= \{\tfrac{1}{2}(k+1)[(k+1)+1]\}^2$$

This is just the RHS of Equation (4.2) with k replaced by $(k+1)$. Hence, the statement is true for $n = k + 1$ and, by induction, is therefore true for all $n \in \mathbb{Z}^+$.

Example 13 Prove that for $n \in \mathbb{Z}^+$, $5^n + 3$ is divisible by 4.

Define
$$f(n) = 5^n + 3.$$

When $n = 1$, $f(1) = 8$, which is divisible by 4, and so the result is true in this case.

Assume the result is true for $n = k$, so that $f(k)$ is a multiple of 4. Then
$$f(k) = 5^k + 3 = N \times 4, \text{ where } N \in \mathbb{Z}^+.$$

Consider now
$$f(k+1) = 5^{k+1} + 3.$$
$$f(k+1) - f(k) = 5^{k+1} - 5^k = 5^k \times (5-1) = 5^k \times 4.$$

Hence,
$$f(k+1) = N \times 4 + 5^k \times 4.$$

The RHS is clearly divisible by 4, and so the result is true for $n = k + 1$. Therefore, by induction, the result is true for all $n \in \mathbb{Z}^+$.

Example 14 Given that $f(n) = n^2 - n$ for $n \in \mathbb{N}$, prove that $f(n)$ is even when $n \geq 2$.

[When $n = 1$, $f(n)$ is zero and therefore there is little point in asking whether it is odd or even. In this problem we take $n_0 = 2$.]

When $n = 2$, $f(2) = 4 - 2 = 2$ and the result is true.
Assume the result is true for $n = k$, so that
$$f(k) = k^2 - k = 2p, \text{ where } p \in \mathbb{Z}^+.$$

Now
$$f(k+1) = (k+1)^2 - (k+1).$$

Therefore,
$$f(k+1) - f(k) = (k+1)^2 - (k+1) - (k^2 - k)$$
$$= k^2 + 2k + 1 - k - 1 - k^2 + k = 2k.$$

Hence,
$$f(k+1) = f(k) + 2k = 2p + 2k,$$
which is clearly even. Therefore, by induction, the result is true for all integral values of $n \geq 2$.

Note, however, that an alternative deductive proof is as follows:
$$f(n) = n(n-1).$$
Therefore $f(n)$ is a product of two consecutive integers, one of which must be even, the other odd. Therefore $f(n)$ is even.

[There may be several ways of proving a result but do not give up if the first method you try does not work.]

Example 15 Given that $n \in \mathbb{N}$ and $n \geq 2$, prove by induction that
$$\left(1 - \frac{1}{2^2}\right)\left(1 - \frac{1}{3^2}\right)\left(1 - \frac{1}{4^2}\right) \cdots \left(1 - \frac{1}{n^2}\right) = \frac{n+1}{2n}.$$

Let us call the LHS T_n. Again we start the induction with $n = 2$. When $n = 2$,
$$T_2 = \left(1 - \frac{1}{2^2}\right) = \frac{3}{4},$$
and the RHS is $\frac{3}{4}$ also. Thus, the result is true for $n = 2$. Assume the result is true for $n = k$, so that
$$T_k = \left(1 - \frac{1}{2^2}\right)\left(1 - \frac{1}{3^2}\right) \cdots \left(1 - \frac{1}{k^2}\right) = \frac{k+1}{2k}. \quad (4.3)$$

When $n = k+1$, the LHS is T_{k+1}, where
$$T_{k+1} = T_k\left[1 - \frac{1}{(1+k)^2}\right].$$

Using Equation (4.3), we may write this as
$$T_{k+1} = \frac{k+1}{2k}\left[1 - \frac{1}{(1+k)^2}\right] = \frac{(k+1)[(k+1)^2 - 1]}{2k(1+k)^2}$$
$$= \frac{(k^2 + 2k)}{2k(k+1)} = \frac{(k+2)}{2(k+1)} = \frac{(k+1) + 1}{2(k+1)}.$$

This is just the RHS of Equation (4.3) with k replaced by $(k+1)$, and so the result is true when $n = k+1$. Hence, by induction, the result is true for all integers $n \geq 2$.

Example 16 The sequence of numbers $u_1, u_2, u_3, \ldots$ is defined by $u_1 = 1$, $u_2 = 5$ and

$$u_{n+2} - 5u_{n+1} + 6u_n = 0, n \geq 1.$$

Prove that $u_n = 3^n - 2^n$.

In this problem we use a generalisation of the method of induction.
Let (S_n) be the statement $(u_n = 3^n - 2^n)$.
(a) When $n = 1$, $3 - 2 = 1 = u_1$.
When $n = 2$, $3^2 - 2^2 = 5 = u_2$.
So (S_n) is true when $n = 1, 2$.
(b) If (S_n) is true when $n = k$ and when $n = k + 1$, then

$$\begin{aligned}u_{k+2} &= 5u_{k+1} - 6u_k \\ &= 5(3^{k+1} - 2^{k+1}) - 6(3^k - 2^k) \\ &= 3^k(5 \times 3 - 6) - 2^k(5 \times 2 - 6) \\ &= 3^k \times 9 - 2^k \times 4 = 3^{k+2} - 2^{k+2}\end{aligned}$$

and so (S_n) is true when $n = k + 2$.
Thus, if (S_n) is true for two consecutive values of n, it is also true for the next value of n. But (S_n) is true for the consecutive values $n = 1, 2$. Hence, by induction, (S_n) is true for all $n \geq 1$.

4.7 Proof of standard results by induction

Arithmetic progression
The standard result is

$$a + (a + d) + \cdots + [a + (n - 1)d] = \tfrac{1}{2}n[2a + (n - 1)d].$$

Let S_n denote the LHS of this equation.
When $n = 1$, $S_1 = a$ and the RHS $= \tfrac{1}{2} \cdot 2a = a$. The stated result is, therefore, true for $n = 1$.
Assume it is true for $n = k$, so that

$$S_k = a + (a + d) + \cdots + [a + (k - 1)d] = \tfrac{1}{2}k[2a + (k - 1)d]. \quad (4.4)$$

Consider

$$\begin{aligned}S_{k+1} &= S_k + (a + kd) \\ &= \tfrac{1}{2}k[2a + (k - 1)d] + (a + kd), \text{ using Equation (4.4)} \\ &= ka + \tfrac{1}{2}k(k - 1)d + a + kd \\ &= (k + 1)a + \tfrac{1}{2}kd(k - 1 + 2) = (k + 1)a + \tfrac{1}{2}kd(k + 1) \\ &= \frac{(k + 1)}{2}[2a + (k + 1 - 1)d],\end{aligned}$$

which is just the RHS of Equation (4.4) with k replaced by $k + 1$. Hence, by induction, the result is true for $n \in \mathbb{Z}^+$.

Geometric progression

The standard result for geometric progressions is

$$a + ar + \cdots + ar^{n-1} = a(1 - r^n)/(1 - r) \quad (r \neq 1).$$

Let S_n denote the LHS of this equation.

When $n = 1$, $S_1 = a$ and the RHS $= [a(1 - r)]/(1 - r) = a$, and so the result is true for $n = 1$.

Assume the result is true for $n = k$, so that

$$S_k = a + ar + \cdots + ar^{k-1} = \frac{a(1 - r^k)}{(1 - r)}. \tag{4.5}$$

It is obvious that

$$S_{k+1} = S_k + ar^k = \frac{a(1 - r^k)}{(1 - r)} + ar^k, \text{ by Equation (4.5)}$$

$$= \frac{a - ar^k + ar^k - ar^{k+1}}{(1 - r)} = a(1 - r^{k+1})/(1 - r).$$

which is just the RHS of Equation (4.5) with k replaced by $k + 1$. Hence, by induction, the result is true for $n \in \mathbb{Z}^+$.

The binomial expansion for $n \in \mathbb{Z}^+$

The standard result we prove by induction is

$$(1 + x)^n = 1 + nx + \frac{n(n - 1)}{2!}x^2 + \cdots$$
$$+ \frac{n(n - 1) \cdots (n - r + 1)}{r!}x^r + \cdots + x^n.$$

It is to be noted that, as we are using induction, n must be a positive integer. Note also that $r! = r(r - 1)(r - 2) \ldots 2.1$. Again let S_n denote the LHS.

When $n = 1$, $S_1 = 1 + x$ and the RHS $= 1 + x + 0 + \cdots + 0$, and so the result is true for $n = 1$.

Assume the result to be true for $n = k$, so that

$$S_k = (1 + x)^k = 1 + kx + \frac{k(k - 1)}{2!}x^2 + \cdots$$
$$+ \frac{k(k - 1) \cdots (k - r + 1)}{r!}x^r + \cdots + x^k. \tag{4.6}$$

Clearly,

$$S_{k+1} = S_k(1 + x)$$

$$= \left[1 + kx + \frac{k(k-1)}{2!}x^2 + \cdots \right.$$
$$\left. + \frac{k(k-1)\cdots(k-r+1)}{r!}x^r + \cdots + x^k\right](1 + x)$$
$$= \left[1 + kx + \frac{k(k-1)}{2!}x^2 + \cdots \right.$$
$$\left. + \frac{k(k+1)\cdots(k-r+1)}{r!}x^r + \cdots + x^k\right]$$
$$+ \left[x + kx^2 + \frac{k(k-1)}{2!}x^3 + \cdots \right.$$
$$\left. + \frac{k(k-1)\cdots(k-r+1)x^{r+1}}{r!} + \cdots + x^{k+1}\right]$$
$$= 1 + (k+1)x + \left[\frac{k(k-1)}{2} + k\right]x^2 + \cdots$$
$$+ A_r x^r + \cdots + x^{k+1},$$

where

$$A_r = \frac{k(k-1)\cdots(k-r+1)}{r!} + \frac{k(k-1)\cdots(k-r+2)}{(r-1)!}$$
$$= \frac{k(k-1)\cdots(k-r+2)}{(r-1)!}\left[\frac{k-r+1}{r} + 1\right]$$
$$= \frac{(k+1)k(k-1)\cdots(k-r+2)}{r!}.$$

This is just the coefficient of x^r in Equation (4.6) with k replaced by $k + 1$. Hence, by induction, the result is true for all $n \in \mathbb{Z}^+$.

Exercise 4

1. Insert the correct symbol $\Rightarrow$ or $\not\Leftarrow$ between the statements
   (a) $(x^2 > 9)$ $(x < -3)$,
   (b) $(x = 2)$ $(x^2 + x - 6 = 0)$.
2. State the negation of the statement

   '$f(x) > x$ for all values of $x > 1$'.

3. Determine which of the following implications are true and which are false:
   (a) $x^2 = 25 \Rightarrow x = 5$.
   (b) $(x - 2)(x + 3) = 0 \Rightarrow x = 2$ or $x = -3$.
   (c) $f(x) = x^2 - 3x + 5 \Rightarrow f(x) > 0$.
4. Using the method of proof by contradiction, show that $\sqrt{2}$ is irrational. (*Hint.* Assume that $\sqrt{2}$ is rational and so can be written in the form p/q.)
5. Find a counter-example to disprove each of the following statements:
   (a) Every number of the form $6n + 1$ is a prime number.
   (b) $a + b > 2\sqrt{(ab)}$.
   (c) $(a - b > 0) \Rightarrow (a^2 - b^2 > 0)$.
6. Prove that $(x^2 - 4x + 3 = 0) \Rightarrow (x = 1$ or $x = 3)$.

7 Use proof by induction to verify the following statements:
   (a) $1^2 + 2^2 + \cdots n^2 = \frac{1}{6}n(n+1)(2n+1)$.
   (b) $3.7^{2n} + 1$ is divisible by $n \in \mathbb{Z}^+$.
   (c) If $u_{n+2} = 10u_{n+1} - 25u_n$ and $u_1 = 0$, $u_2 = 1$, then $u_n = (n-1)5^{n-2}$, $n > 2$.
   (d) $\left(1 - \frac{4}{1}\right)\left(1 - \frac{4}{9}\right)\left(1 - \frac{4}{25}\right) \cdots \left(1 - \frac{4}{(2n-1)^2}\right) = \frac{1+2n}{1-2n}$.
   (e) $2^n > 2n$ for $n \in \mathbb{Z}^+$, $n \geq 2$.

5 Sequences and series

5.1 Sequences
Consider the following sets of numbers

$$2, 4, 6, 8, 10, \qquad (5.1)$$
$$1, 4, 9, 16, 25, \ldots . \qquad (5.2)$$

These are examples of *sequences* of numbers. In any sequence the numbers appear in a given order and, further, there is usually a definite law relating each member to other members. Each member is called a *term* of the sequence.

The sequence (5.1) has just five terms and is an example of a *finite* sequence.

The sequence (5.2) may be written

$$1^2, 2^2, 3^2, 4^2, 5^2, \ldots, \qquad (5.3)$$

where ... means 'and so on without limit'. This sequence has an infinite number of terms and is called an *infinite* sequence.

To define a sequence we require:
(i) the first term,
(ii) the number of terms,
(iii) the law (formula) by which the terms can be calculated.

The first term of a sequence is usually denoted by u_1 and the general term by u_r. Thus, sequence (5.1) is defined by $u_1 = 2$ and $u_r = 2r$. Since this is a finite sequence, the only values of r are 1, 2, 3, 4, 5. We define sequence (5.2) by $u_1 = 1$ and $u_r = r^2$. There is now no restriction on r, so that $r = 1, 2, 3, \ldots$.

Example 1 The general term u_r of a sequence is of the form $u_r = ar + b$, where a and b are constants. Given that $u_1 = 5$ and $u_3 = 11$, find a and b and show that $u_9 = 29$.

$$(u_1 = 5) \Rightarrow (a + b = 5),$$
$$(u_3 = 11) \Rightarrow (3a + b = 11).$$

Solving these simultaneous equations, we obtain

$$(a = 3, \text{ and } b = 2) \Rightarrow (u_r = 3r + 2).$$

Substituting $r = 9$, we obtain $u_9 = 29$.

Example 2 Write down the first five terms of the sequence in which the general term is given by $u_r = 2^r$.

$$u_1 = 2^1 = 2, u_2 = 2^2 = 4, u_3 = 2^3 = 8, u_4 = 2^4 = 16, u_5 = 2^5 = 32.$$

5.2 Series

When the terms of a sequence are added together, we obtain a *series*. For example, sequence (5.1) gives the series

$$2 + 4 + 6 + 8 + 10, \qquad (5.4)$$

which is an example of a *finite series*.

Sequence (5.2) gives the series

$$1 + 4 + 9 + 16 + 25 + \cdots. \qquad (5.5)$$

This is an example of an *infinite series*.

From the general finite sequence $u_1, u_2, \ldots, u_n$ we obtain the series

$$u_1 + u_2 + \cdots + u_n. \qquad (5.6)$$

This series may be written in a more concise form, using what is known as the '*sigma notation*'. Instead of sequence (5.6) we write

$$\sum_{r=1}^{n} u_r. \qquad (5.7)$$

The symbol Σ is a form of the Greek capital letter sigma, which corresponds to S, the first letter of the word 'sum'. In words, expression (5.7) reads 'sigma, r equals 1 to n, of u_r'. The word 'sigma' may be replaced by 'sum'. The expression indicates that a summation is to be carried out, the terms to be added being the u_r, where r is a counter which takes consecutive integral values from 1 to n. The lower limit of the sum is always written below the Σ and the upper limit above. Expression (5.7) is sometimes even further abbreviated to

$$\sum_{1}^{n} u_r.$$

Example 3 Write out explicitly the series

(a) $\sum_{r=1}^{4} \frac{(-1)^r}{r}$, (b) $\sum_{r=0}^{4} (-1)^{r+1} r(r+1)$, (c) $\sum_{r=2}^{4} r!$.

(a) $\sum_{r=1}^{4} \frac{(-1)^r}{r} = \frac{(-1)^1}{1} + \frac{(-1)^2}{2} + \frac{(-1)^3}{3} + \frac{(-1)^4}{4} = -1 + \frac{1}{2} - \frac{1}{3} + \frac{1}{4}.$

(b) $\sum_{r=0}^{4}(-1)^{r+1}r(r+1) = 0 + (-1)^2 1 \times 2 + (-1)^3 2 \times 3 + (-1)^4 3 \times 4$
$\qquad\qquad\qquad\qquad + (-1)^5 4 \times 5$
$\qquad\qquad\qquad = 2 - 6 + 12 - 20.$

(c) $\sum_{r=2}^{4} r! = 2! + 3! + 4!$
$\qquad\quad = 2 + 6 + 24.$

[Remember that $r!$ denotes $r(r-1)(r-2) \ldots 2.1.$]

Note that in (b) and (c) we have $\sum_{r=0}$ and $\sum_{r=2}$, respectively.

Example 4 Write the following series in sigma notation:
(a) $1 - a + a^2 - a^3$,
(b) $\dfrac{1}{2} - \dfrac{1}{4} + \dfrac{1}{8} - \dfrac{1}{16}$.

(a) We note that the general term is of the form $\pm a^r$, with a positive sign when r is even (we regard $r = 0$ as even) and a negative sign when r is odd. The four terms correspond to $r = 0, 1, 2, 3$. Hence, we have

$$1 - a + a^2 - a^3 = \sum_{r=0}^{3}(-1)^r a^r.$$

Check that

$$\sum_{r=1}^{4}(-1)^{r-1}a^{r-1}$$

is an equivalent expression.

(b) We first notice that the series may be written

$$\frac{1}{2} - \frac{1}{2^2} + \frac{1}{2^3} - \frac{1}{2^4}.$$

The general term is now of the form $\pm\dfrac{1}{2^r}$ with a positive sign when r is odd and a negative sign when r is even. The general term can then be written

$$(-1)^{r+1}\frac{1}{2^r}$$

and the series

$$\sum_{r=1}^{4}(-1)^{r+1}\frac{1}{2^r}.$$

Example 5 Write, in sigma notation, the series
$$1.4 + 4.7 + 7.10 + 10.13.$$

Note that the first numbers in the pairs are 1, 4, 7, 10. The difference is 3 in each case and therefore, since the difference is constant, this suggests a linear form such as $ar + b$.

$$r = 1, (ar + b = 1) \Rightarrow (a + b = 1),$$
$$r = 2, (ar + b = 4) \Rightarrow (2a + b = 4),$$
$$\Rightarrow (a = 3, b = -2).$$

Hence, we obtain 1, 4, 7, 10 by substituting the values $r = 1, 2, 3, 4$ in $(3r - 2)$.

The second numbers of the pairs are 4, 7, 10, 13, again differing by 3. Proceeding as above, we find these are obtained from $(3r + 1)$ by substituting the values $r = 1, 2, 3, 4$.

The general term in the series is $(3r - 2)(3r + 1)$ and, hence, the series may be written

$$\sum_{r=1}^{4} (3r - 2)(3r + 1).$$

5.3 Arithmetic progressions (APs)

The sequence 2, 5, 8, 11, ..., 26 is such that each term may be obtained from the previous one by adding a constant, in this case 3. Such a sequence is called an *arithmetic progression*, or AP. In general, if the first term of such a progression is a and a given term differs from the previous one by d, usually called the *common difference*, then the first n terms of the progression are

$$a, (a + d), (a + 2d), \ldots, [a + (n - 1)d]. \qquad (5.8)$$

The common difference d may be positive or negative.

Example 6 The seventh term of an AP is 15 and the tenth term is 21. Find a, the first term of the progression, and d, the common difference. Find also the nth term.

Since the seventh term is 15, $a + 6d = 15$.
Since the tenth term is 21, $a + 9d = 21$.
Solving these equations for a and d, we obtain
$$a = 3, d = 2.$$

The nth term is
$$a + (n - 1)d = 3 + (n - 1)2 = 2n + 1.$$

Example 7 The nth term of an AP is $5 - n$. Find a, the first term of the sequence, and d, the common difference.

The first term is obtained by setting $n = 1$
$$\Rightarrow a = 5 - 1 = 4.$$
The second term is $5 - 2 = 3$ and therefore the common difference d is -1.
An alternative way of proceeding is to write $u_n = 5 - n$.
Then
$$u_{n+1} = 5 - (n + 1).$$
Clearly,
$$u_{n+1} - u_n = \text{common difference } d$$
$$= [5 - (n + 1)] - (5 - n) = -1.$$

Example 8 The eighth term of an AP is five times the second term and the first term is 1. Find the common difference d and the eleventh term.

Since $a = 1$ the eighth term is $1 + 7d$ and the second term is $1 + d$. The given relation between these two terms
$$\Rightarrow [(1 + 7d) = 5(1 + d)]$$
$$\Rightarrow (2d = 4) \Rightarrow (d = 2).$$
The eleventh term is $1 + 10d = 21$.

5.4 Arithmetic series
When the terms of an AP are added together, we obtain an *arithmetic series*. From the sequence (5.8) we obtain the series
$$a + (a + d) + (a + 2d) + \cdots + [a + (n - 1)d]. \quad (5.9)$$
Using the sigma notation, we may write this as
$$\sum_{r=1}^{n} [a + (r - 1)d].$$
[Remember that rth term = first term + $(r - 1) \times$ (common difference)
$$= a + (r - 1)d.]$$
Let us denote the sum of the series (5.9) by S_n so that
$$S_n = a + (a + d) + (a + 2d) + \cdots + [a + (n - 1)d]. \quad (5.10)$$
If we reverse the order of the terms in this series, we also have
$$S_n = [a + (n - 1)d] + [a + (n - 2)d] + \cdots + a. \quad (5.11)$$

Adding Equations (5.10) and (5.11), we obtain

$$2S_n = [2a + (n-1)d] + [2a + (n-1)d] + \cdots + [2a + (n-1)d]$$
$$= n[2a + (n-1)d].$$
$$\Rightarrow S_n = \frac{n}{2}[2a + (n-1)d]. \tag{5.12}$$

If we denote the last term in the series by l, we have $l = a + (n-1)d$. The bracket in Equation (5.12) may then be written as $(a + l)$ and so

$$S_n = \frac{n}{2}(a + l), \tag{5.13}$$

i.e.

$$\tfrac{1}{2}(\text{number of terms}) \times (\text{first term} + \text{last term}).$$

Example 9 Find

$$\sum_{r=1}^{n} r = 1 + 2 + 3 + \cdots + n.$$

In this AP, $a = 1$ and $l = n$.
Hence,

$$S_n = \sum_{r=1}^{n} r = \tfrac{1}{2}(n)(n + 1). \tag{5.14}$$

Example 10 The rth term of an AP is $6r - 1$. Find the sum of n terms of the corresponding arithmetic series.

The first term ($r = 1$) is 5.
The nth term is $6n - 1$.
Hence, by Equation (5.13),

$$S_n = \frac{n}{2}[5 + 6n - 1]$$
$$= \frac{n}{2}[6n + 4] = 3n^2 + 2n.$$

Example 11 The sum of the first n terms of a series is given by $S_n = n^2 - 3n$. Show that the terms of the series are in arithmetic progression. Find a, the first term, and d, the common difference.

It is clear that if T_n is the nth term in the series, then

$$(S_n = S_{n-1} + T_n) \Rightarrow (T_n = S_n - S_{n-1}).$$

Therefore,
$$T_n = n^2 - 3n - [(n-1)^2 - 3(n-1)]$$
$$= n^2 - 3n - [n^2 - 5n + 4]$$
$$= 2n - 4.$$

This is of the form $a - d + nd$ with $d = 2$ and $(a - d = -4) \Rightarrow (a = -2)$.

Example 12 Find the sum of the arithmetic series
$$c + 3c + 5c + \cdots \text{ with 15 terms.}$$

Here the first term $a = c$ and the common difference is $d = 2c$. Hence, by Equation (5.12),
$$S_{15} = \tfrac{15}{2}(2c + 14 \times 2c) = 225c.$$

5.5 Geometric progressions (GPs)

The sequence 3, 6, 12, 24 is such that each term may be obtained from the previous one by multiplying by the same constant, in this case 2. Such a sequence is called a *geometric progression*, or GP. The constant multiplier is called the *common ratio* and is usually denoted by r. In general, if the first term of such a progression is a and the common ratio is r, the first n terms of the progression are

$$a, ar, ar^2, \ldots, ar^{n-1}. \qquad (5.15)$$

The common ratio r may be positive or negative.

Example 13 The fourth term of a GP of real terms is 24 and the seventh term is 192. Find a, the first term, and r, the common ratio. Find also the nth term.

As the fourth term is 24, $ar^3 = 24$.
As the seventh term is 192, $ar^6 = 192$.
Dividing these equations, we obtain
$$(r^3 = 8) \Rightarrow (r = 2).$$
Substituting back, we have
$$(a \times 8 = 24) \Rightarrow (a = 3).$$
The nth term is
$$ar^{n-1} = 3 \times (2)^{n-1}.$$

Example 14 Write down the first three terms of the GP with initial term 3 and common ratio $\tfrac{1}{4}$.

Here $a = 3$ and $r = \frac{1}{4}$.
Hence, the first term is 3, the second term is $3 \times \frac{1}{4} = \frac{3}{4}$ and the third term is $3 \times (\frac{1}{4})^2 = \frac{3}{16}$.

Example 15 Find the first term and the common ratios of the two possible GPs whose third term is 18 and whose fifth term is 162.

With the usual notation:
As the third term is 18, $ar^2 = 18$.
As the fifth term is 162, $ar^4 = 162$.
Dividing, we obtain
$$r^2 = 9 \Rightarrow r = \pm 3.$$
Substituting back, we see that in both cases $a = 2$.
Hence, for one possible GP, $a = 2$ and $r = 3$, and so we have 2, 6, 18, ... For the other possible GP, $a = 2$, $r = -3$, and we have 2, -6, 18, ...

Example 16 Find the number of terms in the GP 2, 1, ..., $\frac{1}{8}$.

Here $a = 2$. As the second term is 1, we have
$$(ar = 1) \Rightarrow (2r = 1) \Rightarrow (r = \tfrac{1}{2}).$$
The nth term of the sequence is given by
$$ar^{n-1} = 2 \times (\tfrac{1}{2})^{n-1} = (\tfrac{1}{2})^{n-2}.$$
If
$$(\tfrac{1}{2})^{n-2} = \tfrac{1}{8} = (\tfrac{1}{2})^3,$$
we have
$$(n - 2 = 3) \Rightarrow (n = 5).$$

5.6 Geometric series
When the terms of a geometric sequence are added together, we obtain a *geometric series*. (This is also often called a geometric progression.) The geometric series obtained from the geometric sequence in (5.15) is
$$a + ar + ar^2 + \cdots + ar^{n-1}. \qquad (5.16)$$
This may be written, using the sigma notation, as
$$a \sum_{p=1}^{n} r^{p-1}.$$

Let us define
$$S_n = a + ar + ar^2 + \cdots + ar^{n-1}. \tag{5.17}$$
Then
$$rS_n = ar + ar^2 + \cdots + ar^{n-1} + ar^n. \tag{5.18}$$
Subtracting (5.18) from (5.17), we obtain
$$(1 - r)S_n = a(1 - r^n).$$
Hence, provided $r \neq 1$,
$$S_n = \frac{a(1 - r^n)}{(1 - r)}. \tag{5.19}$$

If $r = 1$, the series is just $a + a + \cdots + a$ and $S_n = na$.
When $r > 1$, it is more convenient to write (5.19) in the equivalent form
$$S_n = \frac{a(r^n - 1)}{(r - 1)}, \tag{5.20}$$
where both the numerator and denominator are positive.

Example 17 Find the sum of the first six terms of the GP whose first two terms are 3 and 6.

Here $a = 3$ and $r = 2$. From Equation (5.20),
$$S_6 = \frac{3(2^6 - 1)}{2 - 1} = 189.$$

Example 18 Find the sum to n terms of the series
(a) $x + x^2 + x^3 + \cdots, x \neq 1$,
(b) $1 - x + x^2 - x^3 + \cdots, x \neq -1$.

(a) This is a GP with $a = x$ and $r = x$. The nth term is x^n. Using Equation (5.20),
$$S_n = \frac{x(x^n - 1)}{x - 1} = \frac{x^{n+1} - x}{x - 1}.$$

(b) This is a GP with $a = 1$ and $r = -x$.
The nth term is $(-1)^{n-1}x^{n-1}$. Using Equation (5.20),
$$S_n = \frac{1 + (-1)^{n+1}x^n}{1 + x}.$$

62 *Methods of Algebra*

Example 19 Find the sum of the geometric series $1 + 4 + 16 + \cdots + 1024$.

The terms of the series are in GP with $a = 1$ and $r = 4$.
The nth term is $ar^{n-1} = 4^{n-1}$.
If this is 1024, then

$$(4^{n-1} = 1024 = 4^5)$$
$$\Rightarrow (n - 1 = 5)$$
$$\Rightarrow (n = 6).$$

Hence, the required sum is

$$S_6 = \frac{1 \times (4^6 - 1)}{4 - 1} = \frac{4095}{3} = 1365.$$

5.7 Sum of an infinite geometric series

We have seen that the sum of n terms of the general GP is given by Equation (5.19),

$$S_n = \frac{a(1 - r^n)}{(1 - r)}.$$

This may be written

$$S_n = \frac{a}{1 - r} - \frac{ar^n}{1 - r},$$

where the first term is independent of n. An important question is: What happens to the sum S_n as n increases without limit?

If $|r| < 1$, i.e. $-1 < r < 1$, then, as $n \to \infty$, $r^n \to 0$ and therefore $\left|\frac{ar^n}{1 - r}\right|$ can be made less than any positive number ε, however small, by taking n sufficiently large. We denote the sum in this limit by $\lim_{n \to \infty} S_n$, if it exists. We call it the *sum to infinity* and usually write S_∞. The above implies that S_∞ does exist for $|r| < 1$ and that

$$S_\infty = \frac{a}{1 - r}. \qquad (5.21)$$

We say that the series

$$a + ar^2 + ar^3 + \cdots$$

converges to $S_\infty = \frac{a}{1 - r}$ when $|r| < 1$.

If $|r| > 1$, $|r|^n$ increases without limit as n increases and so

$\lim_{n\to\infty} S_n$ does not exist. In this case, we say the series does not converge but *diverges*.

If $|r| = 1$, the terms do not diminish as $n \to \infty$ and so, again, $\lim_{n\to\infty} S_n$ does not exist, i.e. the series does not converge.

Example 20 Find the sum to infinity of the geometric series

$$1 - \frac{1}{4} + \frac{1}{16} - \cdots.$$

For this series, $a = 1$ and $r = -\frac{1}{4}$. Since $|r| < 1$, the sum to infinity exists and is given, by Equation (5.21), as

$$S_\infty = \frac{1}{1 + \frac{1}{4}} = \frac{4}{5}.$$

Example 21 The sum to infinity of a GP is 4 times the first term. Find the common ratio r.

From Equation (5.21),

$$S_\infty = \frac{a}{1 - r}.$$

Since $S_\infty = 4a$, we obtain

$$\left(\frac{a}{1-r} = 4a\right) \Rightarrow \left(1 - r = \frac{1}{4}\right) \Rightarrow \left(r = \frac{3}{4}\right).$$

Example 22 Find the sum to infinity of the geometric series

$$a - \frac{a^2}{b} + \frac{a^3}{b^2} - \cdots, \quad \left|\frac{a}{b}\right| < 1.$$

From Equation (5.21),

$$S_\infty = \frac{a}{1 + a/b},$$

since the first term is a and the common ratio is $-a/b$. Hence,

$$S_\infty = \frac{ab}{b + a}.$$

Example 23 For what range of values of x does the geometric series $1 + 2x + 4x^2 + 8x^3 + \cdots$ converge?

The common ratio is $2x$ and the series only converges for
$$(|2x| < 1) \Rightarrow (|x| < \tfrac{1}{2}) \Rightarrow (-\tfrac{1}{2} < x < \tfrac{1}{2}).$$

5.8 The binomial series

It is easy to show that
$$(1 + x)^2 = 1 + 2x + x^2.$$
Multiplying again by $(1 + x)$ successively, we obtain
$$(1 + x)^3 = 1 + 3x + 3x^2 + x^3,$$
$$(1 + x)^4 = 1 + 4x + 6x^2 + 4x^3 + x^4,$$
$$(1 + x)^5 = 1 + 5x + 10x^2 + 10x^3 + 5x^4 + x^5.$$
If we also recall that
$$(1 + x)^0 = 1 \quad \text{and} \quad (1 + x)^1 = 1 + x,$$
we see that there is a pattern to the coefficients. This is most conveniently displayed in Pascal's triangle (Fig. 5.1).

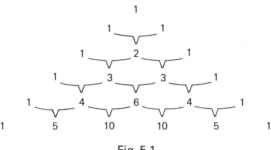

Fig. 5.1

Study of the triangle suggests that
$$(1 + x)^n = 1 + nx + \frac{n(n-1)}{2!}x^2 + \frac{n(n-1)(n-2)}{3!}x^3 + \cdots$$
$$+ \frac{n(n-1) \cdots (n-r+1)}{r!}x^r + \cdots + x^n, \tag{5.22}$$

when $n \in \mathbb{N}$, i.e. n is a positive integer. That this is true was proved by induction in the previous chapter (p. 51).

Using Equation (5.22), we can obtain the general binomial expansion of $(a + x)^n$, where $n \in \mathbb{N}$:
$$(a + x)^n = \left[a\left(1 + \frac{x}{a}\right)\right]^n = a^n\left(1 + \frac{x}{a}\right)^n$$

$$= a^n \left[1 + n\left(\frac{x}{a}\right) + \cdots + \frac{n(n-1)\cdots(n-r+1)}{r!}\left(\frac{x}{a}\right)^r \right.$$
$$\left. + \cdots + \left(\frac{x}{a}\right)^n \right].$$

Hence,

$$(a+x)^n = a^n + na^{n-1}x + \frac{n(n-1)}{2!}a^{n-2}x^2 + \cdots$$
$$+ \frac{n(n-1)\cdots(n-r+1)}{r!}a^{n-r}x^r + \cdots + x^n \quad (5.23)$$

for $n \in \mathbb{N}$.

The expression

$$\frac{n(n-1)\cdots(n-r+1)}{r!}$$

is usually denoted by $\binom{n}{r}$, but sometimes by ^nC_r or $_nC_r$. Hence,

$$\binom{n}{1} = n, \quad \binom{n}{2} = \frac{n(n-1)}{2!} \quad \text{and} \quad \binom{n}{3} = \frac{n(n-1)(n-2)}{3!}.$$

Multiplying the numerator and denominator by $(n-r)!$, we also have

$$\binom{n}{r} = \frac{n!}{r!(n-r)!},$$

a symmetrical form which can often be useful.

It is interesting to enquire what happens when n is not a positive integer. In this case the series

$$1 + nx + \frac{n(n-1)}{2!}x^2 + \frac{n(n-1)(n-2)}{3!}x^3 + \cdots, \quad (5.24)$$

known as the *binomial series*, does not terminate. It can be shown that
(i) the infinite binomial series (5.24) has a limiting sum when $|x| < 1$ but not when $|x| > 1$,
(ii) when the limiting sum exists, its value is $(1+x)^n$.

Example 24 Expand $\left(x + \dfrac{1}{x}\right)^5$.

We use Equation (5.23) with $a = \dfrac{1}{x}$ and $n = 5$.

$$\left(x + \frac{1}{x}\right)^5 = x^5 + 5x^4 \cdot \frac{1}{x} + \frac{5.4}{1.2}x^3 \cdot \frac{1}{x^2} + \frac{5.4.3}{1.2.3}x^2 \cdot \frac{1}{x^3}$$
$$+ \frac{5.4.3.2}{1.2.3.4}x \cdot \frac{1}{x^4} + \frac{1}{x^5}$$
$$= x^5 + 5x^3 + 10x + \frac{10}{x} + \frac{5}{x^3} + \frac{1}{x^5}.$$

Example 25 Calculate the coefficient of x^4 in the expansion of $(2x + 5)^6$.

We write
$$(2x + 5)^6 = 5^6 \left[1 + \frac{2x}{5}\right]^6.$$

The term containing x^4 may then be obtained, using either Pascal's triangle or Equation (5.22). We obtain
$$5^6 \times 15 \times \left(\frac{2x}{5}\right)^4 = 6000x^4.$$

Example 26 Expand $(1 - x)^{1/2}$ as a series of ascending powers of x up to and including the term in x^3.

From (5.24) we have
$$(1 - x)^{1/2} = 1 - \frac{1}{2}x + \frac{(\frac{1}{2})(-\frac{1}{2})}{2!}(-x)^2 + \frac{(\frac{1}{2})(-\frac{1}{2})(-\frac{3}{2})}{3!}(-x)^3 + \cdots$$
$$= 1 - \frac{1}{2}x - \frac{x^2}{8} - \frac{x^3}{16} - \cdots.$$

Example 27 Find the coefficient of x^n in the expansion of $(1 + 2x)^{-3}$.

We write down the first few terms of the binomial series and then see whether we can detect a pattern which will enable us to write down the general term. Using (5.24), we have

$$(1 + 2x)^{-3} = 1 - 3(2x) + \frac{(-3)(-4)}{2!}(2x)^2$$
$$+ \frac{(-3)(-4)(-5)}{3!}(2x)^3 + \cdots + \frac{(-3)(-4)\cdots[-(n+2)]}{n!}(2x)^n +$$

The coefficient of x^2 is
$$\frac{3.4}{2!} \cdot 2^2 = 3.4.2^1.$$

The coefficient of x^3 is
$$(-1)\frac{3.4.5}{3!}.2^3 = (-1)4.5.2^2.$$

The coefficient of x^n is
$$(-1)^n\frac{3.4.5 \ldots (n+2)}{n!}2^n.$$

If we multiply numerator and denominator by 2, we have
$$(-1)^n\frac{(n+2)!}{2.n!}2^n = (-1)^n(n+2)(n+1)2^{n-1}.$$

Example 28 Show that, if x is small compared with unity, so that terms in x^3 and higher powers may be neglected, then
$$\sqrt{\left(\frac{2-x}{2+x}\right)} = 1 - \frac{1}{2}x + \frac{x^2}{8}.$$

We write
$$\sqrt{\left(\frac{2-x}{2+x}\right)} = \sqrt{\left[\frac{(1-x/2)}{(1+x/2)}\right]} = (1-x/2)^{1/2}(1+x/2)^{-1/2}.$$

Using (5.24),
$$\left(1-\frac{x}{2}\right)^{1/2} = 1 - \frac{1}{2}\left(\frac{x}{2}\right) + \frac{\frac{1}{2}(-\frac{1}{2})}{2!}\left(-\frac{x}{2}\right)^2 - \cdots = 1 - \frac{x}{4} - \frac{x^2}{32} - \cdots$$

and
$$\left(1+\frac{x}{2}\right)^{-1/2} = 1 - \frac{1}{2}\left(\frac{x}{2}\right) + \frac{(-\frac{1}{2})(-\frac{3}{2})}{2!}\left(\frac{x}{2}\right)^2 + \cdots = 1 - \frac{x}{4} + \frac{3x^2}{32} + \cdots$$

$$\Rightarrow \sqrt{\left(\frac{2-x}{2+x}\right)} = \left(1 - \frac{x}{4} - \frac{x^2}{32} + \cdots\right)\left(1 - \frac{x}{4} + \frac{3x^2}{32} + \cdots\right)$$
$$= 1 - \frac{x}{2} + \frac{x^2}{8} + \cdots.$$

Alternatively
$$\sqrt{\left(\frac{2-x}{2+x}\right)} = \frac{2-x}{\sqrt{(4-x^2)}}$$

and we may obtain the result by expanding the denominator.
The above expansions are only valid for $|x/2| < 1$, i.e. $|x| < 2$.

Example 29 If x is so small that x^2 and higher powers may be neglected, show that

$$\frac{1}{(x-1)(x+2)} = -\frac{1}{2} - \frac{1}{4}x.$$

We first use partial fractions to write the given expression as the sum of two terms $a/(x-1)$ and $b/(x+2)$. Proceeding as in Chapter 2, we find

$$\frac{1}{(x-1)(x+2)} = \frac{1}{3}\left[\frac{1}{x-1} - \frac{1}{x+2}\right].$$

Now

$$\frac{1}{x-1} = \frac{-1}{1-x} = -1(1-x)^{-1} = -(1 + x + \cdots),$$

using (5.24).
Also,

$$\frac{1}{x+2} = \frac{1}{2(1+x/2)} = \frac{1}{2}\left(1 + \frac{x}{2}\right)^{-1} = \frac{1}{2}\left(1 - \frac{x}{2} + \cdots\right),$$

using (5.24).
Hence,

$$\frac{1}{(x-1)(x+2)} = \frac{1}{3}\left[-(1 + x + \cdots) - \frac{1}{2}\left(1 - \frac{x}{2} + \cdots\right)\right]$$

$$= \frac{1}{3}\left[-\frac{3}{2} - \frac{3}{4}x + \cdots\right] = -\frac{1}{2} - \frac{1}{4}x$$

when terms in x^2 are neglected.

Example 30 Expand $\sqrt{(3 + x^2)}$ in *descending* powers of x as far as the term in $1/x^3$. State the range of values of x for which the expansion is valid.

$$\sqrt{(3 + x^2)} = x\left(1 + \frac{3}{x^2}\right)^{1/2}$$

$$= x\left[1 + \frac{1}{2} \cdot \frac{3}{x^2} + \frac{\frac{1}{2}(-\frac{1}{2})}{1 \cdot 2}\left(\frac{3}{x^2}\right)^2 + \cdots\right]$$

$$= x\left[1 + \frac{3}{2x^2} - \frac{9}{8x^4} + \cdots\right]$$

$$= x + \frac{3}{2x} - \frac{9}{8x^3} + \cdots.$$

The expansion is valid for

$$\left|\frac{3}{x^2}\right| < 1.$$

That is, for

$$|x^2| > 3 \Rightarrow (x > \sqrt{3}) \text{ or } (x < -\sqrt{3}).$$

5.9 Some other finite series

The sum of the first n natural numbers,

$$1 + 2 + 3 + \cdots + n = \sum_{r=1}^{n} r$$

has already been determined as the sum of an arithmetic series. It was found to be $\frac{1}{2}n(n + 1)$. There are other finite sums, involving, for example, powers of the natural numbers, which are of interest.

A very useful method for finding such sums is known as the *method of differences* and uses the following observation. Suppose we wish to find $S_n = \sum_{r=1}^{n} u_r$ and $u_r = f(r + 1) - f(r)$, where f is some function. Then

$$S_n = [f(2) - f(1)] + [f(3) - f(2)] + \cdots + [f(n + 1) - f(n)].$$

A considerable amount of cancellation takes place and we obtain

$$S_n = f(n + 1) - f(1).$$

Example 31 Suppose $f(r) = r^2$.
Then

$$\sum_{r=1}^{n} [(r + 1)^2 - r^2] = (n + 1)^2 - 1.$$

Since

$$(r + 1)^2 - r^2 \equiv r^2 + 2r + 1 - r^2 \equiv 2r + 1,$$

we have

$$\sum_{r=1}^{n} [2r + 1] = (n + 1)^2 - 1$$

$$\Rightarrow 2\sum_{r=1}^{n} r + n = (n^2 + 2n + 1) - 1$$

$$\Rightarrow \sum_{r=1}^{n} r = \frac{1}{2}(n^2 + n) = \frac{n}{2}(n + 1).$$

Example 32 Suppose $f(r) = r^3$.
Then

$$\sum_{r=1}^{n} [(r + 1)^3 - r^3] = (n + 1)^3 - 1.$$

Since

$$(r + 1)^3 - r^3 \equiv r^3 + 3r^2 + 3r + 1 - r^3 \equiv 3r^2 + 3r + 1,$$

we have

$$\sum_{r=1}^{n} (3r^2 + 3r + 1) = (n + 1)^3 - 1$$

$$\Rightarrow 3\sum_{r=1}^{n} r^2 + 3\sum_{r=1}^{n} r + n = (n + 1)^3 - 1.$$

Using the above result for $\sum_{r=1}^{n} r$, we have

$$3\sum_{r=1}^{n} r^2 = (n + 1)^3 - (n + 1) - \tfrac{3}{2}n(n + 1)$$
$$= (n + 1)[(n + 1)^2 - 1 - \tfrac{3}{2}n] = (n + 1)(n^2 + \tfrac{1}{2}n)$$
$$= (n + 1)\frac{n}{2}(2n + 1)$$

$$\Rightarrow \sum_{r=1}^{n} r^2 = \frac{n}{6}(n + 1)(2n + 1).$$

For completeness we state that

$$\sum_{r=1}^{n} r^3 = \tfrac{1}{4}n^2(n + 1)^2.$$

This is set as an exercise and is obtained by taking $f(r) = r^4$.

The results for $\sum_{r=1}^{n} r$, $\sum_{r=1}^{n} r^2$ and $\sum_{r=1}^{n} r^3$ may be used to obtain $\sum_{r=1}^{n} g(r)$, where $g(r)$ is any cubic polynomial in r.

Example 33 Find $S_9 = \sum_{r=1}^{9} (r^3 + 2r + 1)$.

We may write

$$S_9 = \sum_{r=1}^{9} r^3 + 2\sum_{r=1}^{9} r + \sum_{r=1}^{9} 1.$$

Using the above results,

$$S_9 = \tfrac{1}{4}.9^2.10^2 + 2.\tfrac{9}{2}.10 + 9 = 2124.$$

It is unfortunate that no pattern emerges in the results for $\sum_{r=1}^{n} r^k$, where k is any natural number. However, another sequence of results is obtainable, using $f(r)$, of the forms $(r - 1)r$, $(r - 1)r(r + 1)$ and $(r - 1)r(r + 1)(r + 2)$. We give the first of these as an example, leaving the others to be worked as exercises.

Example 34 By using $u_r = f(r+1) - f(r)$ with $f(r) = (r-1)r$, find $\sum_{r=1}^{n} u_r$ and hence obtain $\sum_{r=1}^{n} r$.

Using the given information,

$$\sum_{r=1}^{n} u_r = \sum_{r=1}^{n} [f(r+1) - f(r)] = f(n+1) - f(1)$$
$$= (n+1)n.$$

However,

$$f(r+1) - f(r) = r(r+1) - (r-1)r$$
$$= r^2 + r - r^2 + r = 2r.$$

Hence,

$$\left[2\sum_{r=1}^{n} r = (n+1)n \right]$$

$$\Rightarrow \left[\sum_{r=1}^{n} r = \frac{n}{2}(n+1) \right].$$

Exercise 5

1 Write down the first four terms of the sequences in which the general term u_r is
   (a) $2r - 3$, (b) $\dfrac{r}{r+1}$, (c) $(-1)^r r^2$.

2 Write out the finite sequence of n terms with the general term u_r when
   (a) $u_r = 2r + 3$, $n = 4$,
   (b) $u_r = \dfrac{1}{2^r}$, $n = 6$,
   (c) $u_r = \dfrac{1}{r(r+2)}$, $n = 3$.

3 Write out explicitly the series
   (a) $\sum_{r=1}^{4} 2r$,
   (b) $\sum_{r=0}^{3} 3^r$,
   (c) $\sum_{r=2}^{5} (-1)^r \dfrac{2}{r}$.

4 Express the following series in sigma notation:
   (a) $1 + 2 + 3 + 4 + 5 + 6$,
   (b) $\frac{1}{3} + \frac{1}{5} + \frac{1}{7} + \frac{1}{9}$,
   (c) $1 - 4 + 9 - 16 + 25$,
   (d) $x + 2x^2 + 3x^3 + 4x^4$.

5 The ninth term of an AP is 15 and the fourth term is 40. Find a, the first term, and d, the common difference.

6 The nth term of an AP is $3n - 4$. Find a, the first term, and d, the common difference.

7 The seventh term of an AP is 4 times the second term and the first term is 2. Find d, the common difference, and the tenth term.

8 Find the sum of the arithmetic series
$$1 + 5 + 9 + 13 + \cdots + 49.$$

9 Evaluate
$$\sum_{r=1}^{n} (2r - 1).$$

10 The sum of the first n terms of a series is given by $3n^2 - 4n$. Show that the terms of the series are in arithmetic progression. Find a, the first term, and d, the common difference, and determine the fourth term.

11 The sixth term of a GP is 8 and the third term is 1. Find a, the first term, and r, the common ratio. Write down the nth term.

12 Write down the first four terms of the GP with initial term 1 and common ratio -2.

13 Given that the common ratio r is negative, find the first term a and the common ratio r of the GP of which the third term is 2 and the seventh term is $\frac{1}{8}$.

14 Find the number of terms in the GP
$$3, -6, \ldots, -96.$$

15 The nth term of a GP is given by $(-\frac{1}{3})^n$. Write down the first and the fourth terms.

16 Find the sum of the first six terms of the geometric series
$$1 + 3 + 9 + \cdots.$$

17 Find the sum to n terms of the geometric series
$$1 - x + x^2 - x^3 + \cdots.$$

18 Find the sum to infinity of the GPs
 (a) $1 + \frac{2}{3} + \frac{4}{9} + \cdots$,
 (b) $1 - x + x^2 - \cdots$.

19 The sum to infinity of a GP is 5 times its first term. Find the common ratio.

20 Expand $\left(x^2 - \dfrac{2}{x}\right)^4$.

21 Obtain the coefficient of x^3 in the expansion of $(3 - 2x)^4$.

22 Expand $\dfrac{1}{\sqrt[3]{(8 + x)}}$ as a series of ascending powers of x up to and including the term in x^2.

23 Find the coefficient of x^n in the expansion of $\dfrac{1 + 2x}{1 - 2x}$ as a series of ascending powers of x.

24 Given that x is so small that x^4 and higher powers may be neglected, show that
$$\frac{1}{(1 + 3x)(1 - 2x)} = 1 - x + 7x^2 - 13x^3.$$

For what range of values of x is the expansion valid?

25 Show that, if x is small compared with unity, so that terms in x^3 and higher powers may be neglected,

$$\sqrt{\left(\frac{1+x}{1-x}\right)} = 1 + x + \tfrac{1}{2}x^2.$$

For what range of values of x is the expansion valid?

26 Use the method of differences with $f(r) = r^4$ to show that

$$\sum_{r=1}^{n} r^3 = \tfrac{1}{4}n^2(n+1)^2.$$

27 Show that

$$\sum_{r=1}^{n} (r^3 - r) = \tfrac{1}{4}n(n+1)(n+2)(n-1).$$

28 Use the method of differences with

$$f(r) = (r-1)r(r+1)$$

to show that

$$\sum_{r=1}^{n} r(r+1) = \tfrac{1}{3}n(n+1)(n+2).$$

29 Use the method of differences with

$$f(r) = (r-1)r(r+1)(r+2)$$

to show that

$$\sum_{r=1}^{n} r(r+1)(r+2) = \tfrac{1}{4}n(n+1)(n+2)(n+3).$$

6 Inequalities*

6.1 Linear inequalities
The inequality 'x is greater than y', written $x > y$, is defined to mean $x - y$ is positive. In a similar way, 'x is less than y', written $x < y$, means $x - y$ is negative.
$$(x > y) \Leftrightarrow (x - y > 0).$$
$$(x < y) \Leftrightarrow (x - y < 0).$$

If x is greater than or equal to y, then we write $x \geq y$ with a similar notation, $x \leq y$, for x less than or equal to y.

The basic rules for manipulating inequalities are:

(1) The same number may be added to both sides of an inequality, so that
$$(x > y) \Rightarrow (x + a > y + a).$$

(2) The same number may be subtracted from both sides of an inequality, so that
$$(x > y) \Rightarrow (x - b > y - b).$$

(3) If both sides of an inequality are multiplied by a *positive* number, the inequality is *preserved*.
$$(x > y \text{ and } a > 0) \Rightarrow (ax > ay).$$

(4) If both sides of an inequality are multiplied by a *negative* number the inequality is *reversed*.
$$(x > y \text{ and } b < 0) \Rightarrow (bx < by).$$

(5) The corresponding sides of inequalities of the same kind may be added (but not subtracted).
$$(a > b \text{ and } x > y) \Rightarrow (a + x > b + y).$$

(6) Inequalities of the same type are transitive.
$$(x > y \text{ and } y > z) \Rightarrow (x > z).$$

---

*Throughout this chapter all variables are real (i.e. $\in \mathbb{R}$).

Example 1 Given that $x > y$, show that $x + a > y + a$.

By definition $(x > y) \Leftrightarrow (x - y$ is positive).
We may write $x - y = (x + a) - (y + a)$.

$$[(x + a) - (y + a)] \text{ is positive} \Leftrightarrow (x + a > y + a)$$

Hence,
$$(x > y) \Leftrightarrow (x + a > y + a).$$

Example 2 Given that $x > y$ and $b < 0$, show that $bx < by$.

As above, $(x > y) \Leftrightarrow (x - y$ is positive).
Since $b < 0$, $b(x - y)$ is negative. But $b(x - y) = bx - by$. Hence,

$$(bx - by \text{ is negative}) \Leftrightarrow (bx < by).$$

Example 3 Given that $x > y$ and $a > b$, show that $a + x > b + y$.

$$(x > y) \Leftrightarrow (x - y \text{ is positive}).$$
$$(a > b) \Leftrightarrow (a - b \text{ is positive}).$$

Hence,

$$[(x - y) + (a - b) \text{ is positive}]$$
$$\Rightarrow [(x + a) - (y + b) \text{ is positive}].$$
$$\Rightarrow (x + a > y + b).$$

Example 4 Given that $a > b$, what can be said about the relation between a^2 and b^2?

We need to consider three separate cases depending on the signs of a and b.
(a) a, b both positive.

$$(a > b) \Rightarrow (a^2 > ab), \text{ multiplying by } a,$$
$$(a > b) \Rightarrow (ab > b^2), \text{ multiplying by } b.$$

Combining these, we have

$$(a^2 > ab \text{ and } ab > b^2) \Rightarrow (a^2 > b^2).$$

(b) a, b both negative.

$$(a > b) \Rightarrow (a^2 < ab), \text{ multiplying by } a, \text{ which is negative,}$$
$$(a > b) \Rightarrow (ab < b^2), \text{ multiplying by } b, \text{ which is negative.}$$

Combining these, we have

$$(a^2 < ab < b^2) \Rightarrow (a^2 < b^2).$$

(c) Nothing can be said in the case when a is positive and b is negative. For example,
$$4 > -2 \text{ and } 4^2 > (-2)^2$$
but
$$4 > -6 \text{ and } 4^2 < (-6)^2.$$

Example 5 Find the set of values of x for which $2x + 2 > 6$.

$(2x + 2 > 6) \Rightarrow (2x > 6 - 2 = 4)$, subtracting 2 from each side,
$$\Rightarrow x > 2, \text{ multiplying each side by } \tfrac{1}{2}.$$

The required set is
$$\{x: x > 2\}.$$

Example 6 Find the set of values of x for which
$$\frac{3 + 4x}{x} < 3.$$

(a) When $x > 0$,
$$\left(\frac{3 + 4x}{x} < 3\right) \Rightarrow (3 + 4x < 3x)$$
$$\Rightarrow (4x - 3x < -3) \Rightarrow (x < -3).$$
As this contradicts the condition $x > 0$, the required set cannot contain positive values.

(b) When $x < 0$,
$$\left(\frac{3 + 4x}{x} < 3\right) \Rightarrow (3 + 4x > 3x),$$
since multiplying by a negative number reverses the inequality.
$$(3 + 4x > 3x) \Rightarrow (x > -3).$$
Hence, the required set is
$$\{x: -3 < x < 0\}.$$

6.2 Quadratic inequalities

An inequality of the form $ax^2 + bx + c \geq 0$, where $a \neq 0$, is called a quadratic inequality. The solution depends on the sign of the discriminant $(b^2 - 4ac)$.
(i) If $b^2 \leq 4ac$, then:
$$ax^2 + bx + c \geq 0 \text{ for all values of } x \text{ when } a > 0,$$

and
$$ax^2 + bx + c \leq 0 \text{ for all values of } x \text{ when } a < 0.$$

These results follow from the work of chapter 3.

(ii) If $b^2 > 4ac$, then the equation $ax^2 + bx + c = 0$ has real and distinct roots. If these are α and β, $\beta > \alpha$, then
$$ax^2 + bx + c = a(x - \alpha)(x - \beta).$$

It follows that

	$x - \alpha$	$x - \beta$	$(x - \alpha)(x - \beta)$
$x < \alpha$	$-$ve	$-$ve	$+$ve
$\alpha < x < \beta$	$+$ve	$-$ve	$-$ve
$x > \beta$	$+$ve	$+$ve	$+$ve

(a) For $a > 0$,
$$ax^2 + bx + c > 0 \text{ when } x < \alpha \text{ and when } x > \beta,$$
$$ax^2 + bx + c < 0 \text{ when } \alpha < x < \beta.$$

(b) For $a < 0$,
$$ax^2 + bx + c > 0 \text{ when } \alpha < x < \beta,$$
$$ax^2 + bx + c < 0 \text{ when } x < \alpha \text{ and when } x > \beta.$$

The above results are easily obtained by considering the graph of $y = ax^2 + bx + c$. The two cases are shown in Fig. 6.1 (a, b).

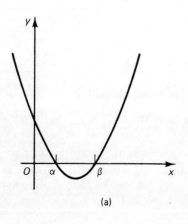
(a)

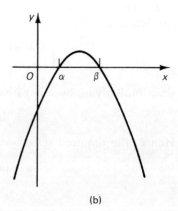
(b)

Fig. 6.1

Example 7 Find the set of values of x for which $2x^2 - 3x - 2 > 0$.

Here, in the usual notation, $a = 2$, $b = -3$, $c = -2$, so that $b^2 > 4ac$. The

LHS may be factorised and we have $(2x + 1)(x - 2) > 0$. In order for the LHS to be positive, $(2x + 1)$ and $(x - 2)$ must have the same sign.
(a) $[(2x + 1) > 0$ and $(x - 2) > 0]$
$$\Rightarrow (x > -\tfrac{1}{2} \text{ and } x > 2).$$
In this case $x > 2$ is the condition required.
(b) $[(2x + 1) < 0$ and $(x - 2) < 0]$
$$\Rightarrow (x < -\tfrac{1}{2} \text{ and } x < 2).$$
In this case the condition is, therefore, $x < -\tfrac{1}{2}$.

The required set is then
$$\{x: x < -\tfrac{1}{2}\} \cup \{x: x > 2\}.$$

If we sketch the curve $y = (2x + 1)(x - 2)$ (see Fig. 6.2), it is easy to see that the marked region gives the same solution set.

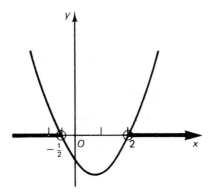

Fig. 6.2

Example 8 Find the set of values for which
$$\frac{3x + 1}{x - 1} < 2.$$

If we multiply this inequality by $(x - 1)$ to eliminate the denominator, we must consider separately the two cases $(x - 1) > 0$ and $(x - 1) < 0$. It is better in this situation to multiply by $(x - 1)^2$, which is always positive provided $x \neq 1$. When $x = 1$, the LHS is not defined.
On multiplying by $(x - 1)^2$, we obtain
$$[(3x + 1)(x - 1) < 2(x - 1)^2]$$
$$\Rightarrow [3x^2 - 2x - 1 < 2x^2 - 4x + 2]$$
$$\Rightarrow [x^2 + 2x - 3 < 0]$$
$$\Rightarrow [(x + 3)(x - 1) < 0].$$

For this to be satisfied, $(x + 3)$ and $(x - 1)$ must have different signs.
(a) $[(x + 3) > 0$ and $(x - 1) < 0]$
$$\Rightarrow [x > -3 \text{ and } x < 1]$$
$$\Rightarrow (-3 < x < 1).$$
(b) $[(x + 3) < 0$ and $(x - 1) > 0]$
$$\Rightarrow (x < -3 \text{ and } x > 1).$$

There are no values of x which satisfy these last two conditions. Hence, the required set is
$$\{x: -3 < x < 1\}.$$

The result is also easily deduced from Fig. 6.3, where we have plotted $y = x^2 + 2x - 3$.

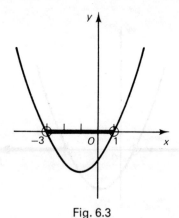

Fig. 6.3

6.3 Inequalities involving the modulus sign

The symbol $|x|$ is called the *modulus* of x. It is defined to be the *magnitude* of x such that

$|x| = x$ when x is positive,
$|x| = -x$ when x is negative.

The definition implies $|x|^2 = x^2$ for all values of x.

Given that $|ax + b| > c$, where $c > 0$, we have, from the above, $[(ax + b)^2 > c^2$ for all $x]$.
$$\Rightarrow [a^2x^2 + 2abx + (b^2 - c^2) > 0].$$

This is now a quadratic inequality and may be solved by the methods described on pp. 77–9.

Example 9 Find the set of values of x for which $|x + 1| > 1$.

$$[|x + 1| > 1] \Rightarrow [|x + 1|^2 > 1] \Rightarrow [(x + 1)^2 > 1]$$
$$\Rightarrow [x^2 + 2x > 0] \Rightarrow [x(x + 2) > 0].$$

(a) $(x > 0$ and $x + 2 > 0)$
$$\Rightarrow (x > 0 \text{ and } x > -2).$$
Therefore, we require $x > 0$.

(b) $(x < 0$ and $x + 2 < 0)$
$$\Rightarrow (x < 0 \text{ and } x < -2).$$
Therefore, we require $x < -2$.

The required set of values is
$$\{x: x < -2\} \cup \{x: x > 0\}.$$

Inequalities involving modulus signs may also be treated graphically. The graph of $y = |x|$ is easily drawn, since $y = x$ when x is positive, $y = -x$ when x is negative (see Fig. 6.4).

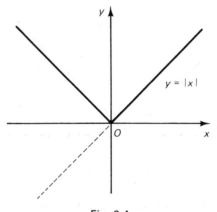

Fig. 6.4

To obtain the graph of $y = |ax + b|$, where $a > 0$, consider first $y = ax + b$. We note that $y = 0$ when $x = -b/a$. Hence,

$$|ax + b| = ax + b \text{ for } x > -b/a.$$
$$|ax + b| = -(ax + b) \text{ for } x < -b/a.$$

The graph of $y = |ax + b|$ for the case $a > 0$, $b > 0$, is shown in Fig. 6.5. Notice that the part to the left of $x = -b/a$ is obtained by reflecting in the x-axis that part of the line $y = ax + b$ which is below the x-axis.

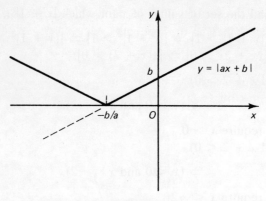

Fig. 6.5

Example 10 Solve graphically $|x + 1| > 1$.

Draw the graphs of $y = |x + 1|$ and $y = 1$ (see Fig. 6.6). These graphs intersect at $x = 0$ and $x = -2$. Clearly the graph of $y = |x + 1|$ is above the graph of $y = 1$ outside the region $-2 \leq x \leq 0$. Hence, the required solution is

$$\{x: x < -2\} \cup \{x: x > 0\}.$$

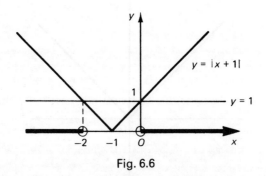

Fig. 6.6

Example 11 Find the set of values of x for which $|4 - 3x| \leq |2x - 1|$.

$$[|4 - 3x| \leq |2x - 1|] \Rightarrow [|4 - 3x|^2 \leq |2x - 1|^2]$$
$$\Rightarrow [(4 - 3x)^2 \leq (2x - 1)^2] \Rightarrow [16 + 9x^2 - 24x \leq 4x^2 + 1 - 4x]$$
$$\Rightarrow [5(x^2 - 4x + 3) \leq 0] \Rightarrow [5(x - 3)(x - 1) \leq 0].$$

(a) $[x - 3 \leq 0$ and $x - 1 \geq 0]$

$$\Rightarrow [x \leq 3 \text{ and } x \geq 1]$$
$$\Rightarrow [\text{required set is } \{x: 1 \leq x \leq 3\}].$$

(b) $[x - 3 \geq 0$ and $x - 1 \leq 0]$

$$\Rightarrow [x \geq 3 \text{ and } x \leq 1].$$

There are no values of x which satisfy both of these conditions.
The set of values satisfying the inequality is

$$\{x: 1 \leq x \leq 3\}.$$

To solve this inequality graphically, we draw the graphs of $y = |4 - 3x|$ and $y = |2x - 1|$. These intersect at $x = 1$ and $x = 3$ (see Fig. 6.7). The graph of

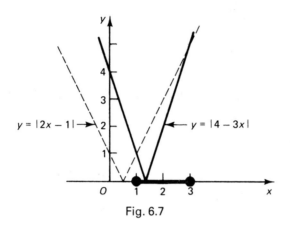

Fig. 6.7

$y = |4 - 3x|$ lies below the graph of $y = |2x - 1|$ between these points. Hence, the required set is, again,

$$\{x: 1 \leq x \leq 3\}.$$

6.4 More general inequalities in one variable

We will here discuss various examples to illustrate other useful techniques for solving inequalities.

Example 12 Find the set of values of x for which $(x - 2)(x - 3)(x - 1) \leq 0$.

Let $f(x) \equiv (x - 2)(x - 3)(x - 1)$. Clearly, $f(x) = 0$ when $x = 1, 2$ or 3. These are often called the *critical points*. We consider the sign of $f(x)$ in the regions between these points and to the left of $x = 1$ and the right of $x = 3$.

	$(x - 1)$	$(x - 2)$	$(x - 3)$	$f(x)$
$x < 1$	−ve	−ve	−ve	−ve
$1 < x < 2$	+ve	−ve	−ve	+ve
$2 < x < 3$	+ve	+ve	−ve	−ve
$x > 3$	+ve	+ve	+ve	+ve

We deduce that $f(x) \leq 0$ for $x \leq 1$ and $2 \leq x \leq 3$. The solution set is

$$\{x: x \leq 1\} \cup \{x: 2 \leq x \leq 3\}.$$

Example 13 Find the set of values of x for which
$$\frac{x^2 - 2}{x - 3} < 2.$$

When solving an inequality such as $f(x) \leq g(x)$, it is usually good policy to rearrange it in the equivalent form $f(x) - g(x) \leq 0$ and then, if possible, as $h(x) \leq 0$.

$$\left[\frac{x^2 - 2}{x - 3} < 2\right] \Rightarrow \left[\frac{x^2 - 2}{x - 3} - 2 < 0\right].$$

The LHS may now be written as
$$\frac{x^2 - 2 - 2(x - 3)}{x - 3} = \frac{x^2 - 2x + 4}{x - 3}$$

so that we now have
$$\frac{x^2 - 2x + 4}{x - 3} < 0.$$

As before, we multiply by $(x - 3)^2$, which is positive when $x \neq 3$. (The LHS is not defined when $x = 3$.) We have
$$(x^2 - 2x + 4)(x - 3) < 0$$

From the results of Section 6.2, $x^2 - 2x + 4 \geq 0$ for all x. Hence, we require $x - 3 < 0$. That is, the solution set is
$$\{x: x < 3\}.$$

Example 14 Find the set of values of x for which
$$\frac{x - 1}{x + 1} \leq x.$$

$$\left[\frac{x - 1}{x + 1} \leq x\right] \Rightarrow \left[\frac{x - 1}{x + 1} - x \leq 0\right].$$

Multiplying by $(x + 1)^2$, which is positive $(x \neq -1)$,
$$[(x + 1)(x - 1) - x(x + 1)^2 \leq 0]$$
$$\Rightarrow [(x + 1)[x - 1 - x(x + 1)] \leq 0]$$
$$\Rightarrow [(x + 1)[-1](1 + x^2) \leq 0].$$

As above, $1 + x^2$ is always positive, so we require
$$[(x + 1) > 0] \Rightarrow (x > -1).$$

The solution set is
$$\{x: x > -1\}.$$

Example 15 Find the set of values of x for which
$$|2x + 3| - |x + 4| < 2.$$

$2x + 3 = 0 \Rightarrow x = -\frac{3}{2}$
and so
$$|2x + 3| = 2x + 3 \quad \text{for } x \geq -\frac{3}{2},$$
$$|2x + 3| = -(2x + 3) \quad \text{for } x \leq -\frac{3}{2}.$$

$x + 4 = 0 \Rightarrow x = -4$
and so
$$|x + 4| = x + 4 \quad \text{for } x \geq -4,$$
$$|x + 4| = -(x + 4) \quad \text{for } x \leq -4.$$

We now consider the regions in which the given inequality takes different forms.

(a) $x \leq -4$.
The inequality now takes the form
$$[-(2x + 3) + (x + 4) < 2]$$
$$\Rightarrow (-x + 1 < 2)$$
$$\Rightarrow x > -1.$$

There are no values of x satisfying both $x \leq -4$ and $x > -1$.

(b) $-4 \leq x \leq -\frac{3}{2}$.
In this region the inequality becomes
$$[-(2x + 3) - (x + 4) < 2]$$
$$\Rightarrow (-3x - 7 < 2)$$
$$\Rightarrow (3x > -9) \Rightarrow (x > -3).$$

The set of values satisfying $x > -3$ and $-4 \leq x \leq -\frac{3}{2}$ is
$$\{x: -3 < x \leq -\frac{3}{2}\}.$$

(c) $x \geq -\frac{3}{2}$.
$$[(2x + 3) - (x + 4) < 2]$$
$$\Rightarrow (x - 1 < 2) \Rightarrow (x < 3)$$

Hence, the values of x for which $-\frac{3}{2} \leq x < 3$ satisfy the inequality. Combining (b) and (c), we obtain the solution set
$$\{x: -3 < x < 3\}.$$

Example 16 Prove that for real values of x the values of
$$\frac{6x + 5}{3x^2 + 4x + 2}$$

cannot lie outside $-\frac{3}{2}$ to 3.

Let
$$\frac{6x + 5}{3x^2 + 4x + 2} = N.$$

Then
$$[6x + 5 = N(3x^2 + 4x + 2)]$$
$$\Rightarrow [3Nx^2 + x(4N - 6) + (2N - 5) = 0].$$

For a particular value of N, x can be found, provided that
$$[(4N - 6)^2 \geq 4.3N(2N - 5)]$$
$$\Rightarrow [4N^2 + 9 - 12N \geq 6N^2 - 15N]$$
$$\Rightarrow (2N^2 - 3N - 9 \leq 0)$$
$$\Rightarrow [(2N + 3)(N - 3) \leq 0].$$

(a) $[2N + 3 \geq 0$ and $N - 3 \leq 0]$
$$\Rightarrow [N \geq -\tfrac{3}{2} \text{ and } N \leq 3]$$
$$\Rightarrow [-\tfrac{3}{2} \leq N \leq 3].$$

(b) $[2N + 3 \leq 0$ and $N - 3 \geq 0]$
$$\Rightarrow [N \leq -\tfrac{3}{2} \text{ and } N \geq 3].$$

There are no values of N for which $N \geq 3$ and $N \leq -\tfrac{3}{2}$.
Hence, $-\tfrac{3}{2} \leq N \leq 3$ as required.

6.5 Inequalities in two variables

We have seen above that the solution of an inequality in one variable is a set of points on the real line.

The solution of an inequality in two variables x and y of the form $f(x, y) > 0$ is a set of points (x, y) in the x,y-plane. The equation $f(x, y) = 0$ is the equation of a curve C in the x,y-plane which divides the plane into two regions. In general, in one of these regions $f(x, y)$ is greater than 0 and in the other $f(x, y)$ is less than 0. Which region is which is easily determined by finding the sign of $f(x, y)$ for just one point.

Example 17 Determine the region of the x, y-plane for which $x^2 + y^2 > 9$.

We first write the inequality in the form
$$f(x, y) \equiv x^2 + y^2 - 9 > 0.$$

The curve C given by $x^2 + y^2 - 9 = 0$, or $x^2 + y^2 = 9$, is a circle whose centre is the origin and of radius 3.
The origin $(0, 0)$ is inside the circle and
$$f(0, 0) = -9$$

and so is not in the required region. The required region is therefore the set of points outside the circle (see Fig. 6.8).

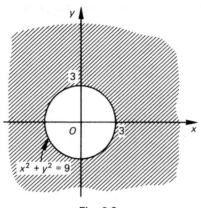

Fig. 6.8

Example 18 Determine the region of the x,y-plane for which $x + y \leq 1$.

Write this as $f(x, y) = x + y - 1 \leq 0$. Since the given inequality includes the $=$ sign, the region will include the curve $x + y = 1$. Let us now seek the region for which $x + y < 1$, i.e. $f(x, y) < 0$.

The curve $x + y = 1$ is of course a straight line. This line divides the plane into two half-planes. Substituting $(0, 0)$, we obtain

$$f(0, 0) = -1 < 0$$

and so $(0, 0)$ is in the required region.

The shaded region, including the line, is therefore the required set of points (see Fig. 6.9).

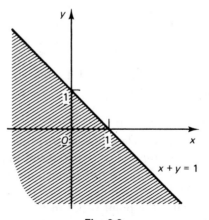

Fig. 6.9

If we add the further constraints that $x \geq 0$ and $y \geq 0$, we obtain the region shown in Fig. 6.10, all the boundaries being included.

When the boundary (or part of it) is included, the relevant parts are usually indicated by heavier curves, as shown in Fig. 6.10.

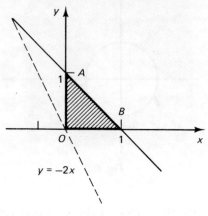

Fig. 6.10

It is quite often necessary to obtain the greatest or least values for points in such a region. (Problems of this nature occur in linear programming.) For example, we might ask: What is the greatest value of $z = 2x + y$ for points satisfying the given inequalities?

The curve $2x + y = k$ is a straight line parallel to the dotted line. As we move this line to the right, k increases. The greatest value will therefore occur when the line is as far from the origin as possible, i.e. when it passes through the point B. The value of z at B is $2(1) + 0 = 2$.

Example 19 Determine the region of the x,y-plane for which

$$(x^2 + y^2 - 9)(y^2 - 4x) > 0.$$

The given inequality can only be satisfied if either
(a) both brackets are positive, or
(b) both brackets are negative.
(a) $x^2 + y^2 - 9 > 0$, $y^2 - 4x > 0$.
The curve C_1: $x^2 + y^2 = 9$ is a circle whose centre is the origin and of radius 3. At the point $(0, 0)$ the function $x^2 + y^2 - 9 < 0$ and so is not in the required region. The required region is the set of points outside the circle.
The curve C_2: $y^2 = 4x$ is a parabola. At the point $(1, 0)$ the function $y^2 - 4x < 0$ and so, again, the point is not in the given region.
Hence, the required region is the region shaded in Fig. 6.11.

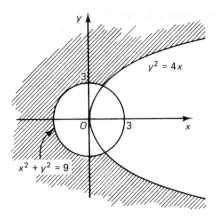

Fig. 6.11

(b) $x^2 + y^2 - 9 < 0$, $y^2 - 4x < 0$.

We may use the above analysis. In this case we require the alternative regions in both cases and so we obtain the region shaded in Fig. 6.12.
The complete picture is obtained by superimposing Fig. 6.11 on Fig. 6.12.

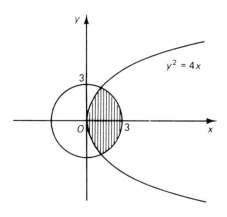

Fig. 6.12

Exercise 6

1. Given that $x > y$, show that $x - b > y - b$.
2. Given that $x > y$ and $a > 0$, show that $ax > ay$.
3. Given that $a > b$ and a and b are positive, show that, for any positive integer n, $a^n > b^n$.
4. Find the set of values of x for which
$$3x - 2 > 7.$$

5 Find the set of values of x for which
$$\frac{2x+3}{x-1} < 1.$$

6 Find the set of values of x for which
$$3x - 2 > x^2.$$

7 Find the set of values of x for which
$$\frac{x-5}{2-x} > 3.$$

8 Find the set of values of x for which
$$|x - 3| < 4.$$

9 Find the set of values of x for which
$$|1 - 2x| \leq |3x - 1|.$$

10 Find the set of values of x for which
$$(x + 2)(x - 2)(x + 7) > 0.$$

11 Find the set of values of x for which
$$\frac{x^2 + 56}{x} > 15.$$

12 Show that for real values of x the values of the function $\dfrac{x^2 + 2}{2x + 1}$ cannot lie between -2 and 1.

13 Determine the region of the x,y-plane for which $x \geq 0$ and $x - y \leq 1$. Hence find the minimum value of y.

14 Shade the region of the x,y-plane for which $y^2 < 4x$, $x - y < \frac{5}{4}$. Hence find the maximum and minimum values of y.

15 Determine the region of the x,y-plane for which
$$(x^2 + y^2 - 4)(y - x^2) < 0.$$

ANSWERS

Exercise 1
1. (a) $x \in \mathbb{R}, y \in \mathbb{R}$;
   (b) $x \in \mathbb{R}, \{y: 0 \leq y \leq 1\}$;
   (c) $x \in \mathbb{R}, \{y: y \geq 2\}$
2. (a) Neither, (b) even, (c) even
3. gf: $x \mapsto (2x - 1)^3$,
   fg: $x \mapsto 2x^3 - 1$,
   ff: $x \mapsto 4x - 3$,
   gg: $x \mapsto x^9$
4. (a) $f^{-1}: x \mapsto \frac{1}{2}(5 - x), x \in \mathbb{R}$
   (b) $g^{-1}: x \mapsto 1 + \frac{4}{x}, x \in \mathbb{R}, x \neq 0$
   (c) $h^{-1}: x \mapsto \frac{1}{2}[\sqrt{x} - 1], x \in \mathbb{R}^+, x = 0$
5. (a) $r_1^{-1}: x \mapsto \frac{1}{4}\sqrt{x}$, domain $x \in \mathbb{R}$, codomain $x \geq 0$
   (b) $r_2^{-1}: x \mapsto \frac{1}{x} - 2$, domain $x \in \mathbb{R}(x \neq -2)$, codomain $x \in \mathbb{R}(x \neq 0)$
   (c) $r_3^{-1}: x \mapsto \sqrt{(x + 4)}$, domain $x \in \mathbb{R}$, codomain $x \geq -4$.
6. Range -8 to 4,
   f has inverse if x restricted to $-2 < x \leq 2$,
   $f^{-1}: x \mapsto \frac{2 + x}{3}, -2 < x \leq 1$,
   $\mapsto \sqrt{x}, 1 < x \leq 2$
7. (a) $\frac{1}{27}$, (b) $\frac{25}{4}$, (c) $\frac{5}{2160}$, (d) 9
   (e) $3\frac{1}{30}$
8. (a) x^5, (b) $y^{-1/3}$, (c) x, (d) $\frac{3}{x}$,
   (e) $\frac{-1 - 2x}{(x + 1)^{2/3}}$, (f) $\frac{x^2 - 3x + 2}{x^3}$
9. (a) $3\sqrt{10}$, (b) $2\sqrt{3}$, (c) 30,
   (d) $4 + 3\sqrt{3}$, (e) 0
10. (a) $\frac{\sqrt{3}}{3}$, (b) $\frac{\sqrt{6} + 2}{2}$, (c) 2
11. (a) $2 = \log_3 9$, (b) $0 = \log_8 1$,
    (c) $-4 = \log_{1/3} 81$, (d) $b = \log_a 2$
12. (a) -3, (b) -1, (c) $\frac{3}{2}$

13. (a) $10^1 = 10$, (b) $3^4 = 81$,
    (c) $27^{1/3} = 3$
14. (a) $\log_a \frac{3}{2}$, (b) 2
16. $\frac{e^x - 1}{e}$
17. (a) $X = u, Y = \ln v, Y = \ln a + nX$,
    (b) $X = t, Y = \frac{s}{t}, Y = u + \frac{1}{2}fX$,
    (c) $X = \ln x, Y = \ln y$,
    $kX + Y = \ln a$

Exercise 2
1. $5x^3 + 3x^2 + 3x + 10$
2. $x^3 - 2x^2 + 11x + 4$
3. $x^5 + x^4 + 3x^3 + 3x^2 + 3x + 1$
4. $x^2 + 3x + 5$
5. Quotient $(x + 4)$, remainder $(5x - 4)$
6. (a) -3, (b) $\frac{23}{8}$
7. $a = 1, b = 2, c = 3$; remainder -5
9. $a = 2, b = -3; (2x + 1)$
11. $x + 2$ is a factor; $(x + 2)(x^2 + x + 4)$
12. $(x - 2), (3x + 2), (x + 3)$
13. (a) $\frac{5x}{(x - 2)(x + 3)}$,
    (b) $\frac{-3x^2 - x}{(x^2 + x + 2)(x + 3)}$
14. (a) $\frac{1}{x - 1} + \frac{1}{x + 3}$,
    (b) $\frac{2x - 3}{x^2 + x + 1} + \frac{1}{x + 1}$,
    (c) $\frac{-1}{x + 2} + \frac{2}{(x + 2)^2} + \frac{1}{(x - 3)}$
15. $(x - 1) + \frac{7/5}{(x - 2)} + \frac{8/5}{(x + 3)}$

Exercise 3
1. (a) $f(0) = 1$, min. at $(-\frac{1}{2}, 0)$ where it touches x-axis

Answers 91

(b) $g(0) = 4$, max. at $(-\frac{3}{2}, 6\frac{1}{4})$, cuts x axis at $x = 1, -4$
  (c) $h(0) = 3$, min. at $(\frac{1}{2}, 2\frac{3}{4})$, does not cut x axis
2 (a) No real solutions,
  (b) 0·56 or −3·56,
  (c) 2 or 3,   (d) $\frac{1}{3}$ twice.
3 (a) $x < 0, x > 1$,
  (b) No real values of x,
  (c) $-1·62 \leq x \leq 0·62$,
  (d) All real values of x
4 (a) $x^2 - 4x + 2 = 0$,
  (b) $2x^2 - 12x + 17 = 0$,
  (c) $x^2 - 12x + 4 = 0$
5 $p = \frac{1}{3}$ or $-1, p < -1$
6 $c = 2$
7 60 m, 15 m
8 (b) $k \leq 0, k \geq 3$;   (c) $k \geq 3$.

Exercise 4
1 (a) $\Leftarrow$,   (b) $\Rightarrow$
2 '$f(x) \leq x$ for at least one value of $x > 1$'
3 (a) false,   (b) true,   (c) true
5 (a) When $n = 4$, the number is 25, which is not prime,
  (b) $a = -1, b = -2$,
  (c) $a = 2, b = -5$

Exercise 5
1 (a) $-1, 1, 3, 5$,
  (b) $\frac{1}{2}, \frac{2}{3}, \frac{3}{4}, \frac{4}{5}$,
  (c) $-1, 4, -9, 16$
2 (a) $5, 7, 9, 11$,
  (b) $\frac{1}{2}, \frac{1}{4}, \frac{1}{8}, \frac{1}{16}, \frac{1}{32}, \frac{1}{64}$,
  (c) $\frac{1}{3}, \frac{1}{8}, \frac{1}{15}$
3 (a) $2 + 4 + 6 + 8$,
  (b) $1 + 3 + 9 + 27$,
  (c) $1 - \frac{2}{3} + \frac{1}{2} - \frac{2}{5}$
4 (a) $\sum_{r=1}^{6} r$,   (b) $\sum_{r=1}^{4} \frac{1}{(2r+1)}$,

  (c) $\sum_{r=1}^{5} (-1)^{r+1} r^2$,   (d) $\sum_{r=1}^{4} rx^r$
5 $a = 55, d = -5$
6 $a = -1, d = 3$
7 $d = 3$, tenth term 29
8 325
9 n^2
10 $a = -1, d = 6$, fourth term 17
11 $a = \frac{1}{4}, r = 2$, nth term 2^{n-3}
12 $1, -2, 4, -8$
13 $a = 8, r = -\frac{1}{2}$
14 6
15 $-\frac{1}{3}, \frac{1}{81}$
16 364
17 $\dfrac{1 + (-1)^{n+1} x^n}{1 + x}$
18 (a) 3,   (b) $\dfrac{1}{1 + x}$
19 $r = \frac{4}{5}$
20 $x^8 - 8x^5 + 24x^2 - \dfrac{32}{x} + \dfrac{16}{x^4}$
21 -96
22 $\frac{1}{2} - \frac{1}{48}x + \frac{1}{576}x^2$
23 2^{n+1}
24 $-\frac{1}{3} < x < \frac{1}{3}$
25 $-1 < x < 1$

Exercise 6
4 $\{x: x > 3\}$
5 $\{x: -4 < x < 1\}$
6 $\{x: 1 < x < 2\}$
7 $\{x: 2 < x < 2\frac{3}{4}\}$
8 $\{x: -1 < x < 7\}$
9 $\{x: x \leq 0\} \cup \{x: x \geq \frac{2}{5}\}$.
10 $\{x: -7 < x < -2\} \cup \{x: x > 2\}$
11 $\{x: 0 < x < 7\} \cup \{x: x > 8\}$
13 $y = -1$
14 $y = 5, y = -1$

Index

algebraic functions 1
arithmetic progression (AP) 50, 57
arithmetic series 58
   sum of 58

binomial expansion 51
binomial series 65

codomain 1
coefficients 17
composite function 3
contradiction, proof by 44
converse 42
counter-example 45
cubic 17

deduction, proof by 45
degree of polynomial 17
dependent variable 1
divisor 19
domain 1

equation 4
even function 2
exhaustion, proof by 45
exponential function 11

factor theorem 21
factors of polynomials 22
functions 1

geometric progression (GP) 51, 60
geometric series 61
   sum of 62
graph 2

identity 4

independent variable 1
indices 6
induction, proof by 46
inequalities 75
   in 1 variable 83
   in 2 variables 86
infinite geometric series 63
inverse function 4

linear inequalities 75
linear relations 12
logarithmic function 11
logarithms 8
   rules of 9
logical concepts 42

method of differences 70
modulus 80
modulus sign in inequalities 80

natural logarithms 11
necessary condition 43
negation 43

odd function 2
order of polynomial 17

partial fractions 24
polynomials 17
   division of 19
   factors of 22
   manipulation of 17
   multiplication of 18
proof 42

range (set) 1
rational functions 24

rationalising 7
remainder theorem 20

sequences 54
  finite 54
  infinite 54
series 55

finite 55
  infinite 55
sigma notation 55
statement 42
sufficient condition 43
sum to infinity 63
surds 7